T0353578

Never Waste a Good Crisis

This book covers statistical consequences of breaches of research integrity such as fabrication and falsification of data, and researcher glitches summarized as questionable research practices. It is unique in that it discusses how unwarranted data manipulation harms research results and that questionable research practices are often caused by researchers' inadequate mastery of the statistical methods and procedures they use for their data analysis. The author's solution to prevent problems concerning the trustworthiness of research results, no matter how they originated, is to publish data in publicly available repositories and encourage researchers not trained as statisticians not to overestimate their statistical skills and resort to professional support from statisticians or methodologists.

The author discusses some of his experiences concerning mutual trust, fear of repercussions, and the bystander effect as conditions limiting revelation of colleagues' possible integrity breaches. He explains why people are unable to mimic real data and why data fabrication using statistical models stills falls short of credibility. Confirmatory and exploratory research and the usefulness of preregistration, and the counter-intuitive nature of statistics, are discussed. The author questions the usefulness of statistical advice concerning frequentist hypothesis testing, Bayes-factor use, alternative statistics education, and reduction of situational disturbances like performance pressure, as stand-alone means to reduce questionable research practices when researchers lack experience with statistics.

Klaas Sijtsma is an emeritus professor of methods and techniques of psychological research at Tilburg University. He is also a former dean of a school suffering from a huge data fraud affair by one of its eminent professors, which he had to cope with during his term. Being both an applied statistician and an experienced administrator having to deal with serious breaches of scientific integrity gave him a unique perspective on the problems discussed in this book. He gave many lectures on the topic and authored several publications. This book summarizes his views. In addition, he has published more than 200 papers and book chapters on statistical topics and coauthored three books on measurement of psychological attributes such as intelligence, personality traits, and attitudes. At present, he is co-chair of the Committee on Research Integrity of Erasmus University Rotterdam.

ASA-CRC Series on
STATISTICAL REASONING IN SCIENCE AND SOCIETY

SERIES EDITORS

Nicholas Fisher,
University of Sydney, Australia

Nicholas Horton,
Amherst College, MA, USA

Regina Nuzzo,
Gallaudet University, Washington, DC, USA

David J Spiegelhalter,
University of Cambridge, UK

PUBLISHED TITLES

Measuring Society
Chaitra H. Nagaraja

Monitoring the Health of Populations by Tracking Disease Outbreaks
Steven E. Fricker and Ronald D. Fricker, Jr.

Debunking Seven Terrorism Myths Using Statistics
Andre Python

Achieving Product Reliability: A Key to Business Success
Necip Doganaksoy, William Q. Meeker, and Gerald J. Hahn

Protecting Your Privacy in a Data-Driven World
Claire McKay Bowen

Backseat Driver: The Role of Data in Great Car Safety Debates
Norma F. Hubele

Statistics Behind the Headlines
A. John Bailer and Rosemary Pennington

Never Waste a Good Crisis: Lessons Learned from Data Fraud and Questionable Research Practices
Klaas Sijtsma

For more information about this series, please visit: https://www.crcpress.com/go/asacrc

Never Waste a Good Crisis

Lessons Learned from Data Fraud and Questionable Research Practices

Klaas Sijtsma

CRC Press
Taylor & Francis Group
Boca Raton London New York

CRC Press is an imprint of the
Taylor & Francis Group, an **informa** business

A CHAPMAN & HALL BOOK

Designed cover image: © Shutterstock, ID 1052140007. Lightspring

First edition published 2023
by CRC Press
6000 Broken Sound Parkway NW, Suite 300, Boca Raton, FL 33487-2742

and by CRC Press
4 Park Square, Milton Park, Abingdon, Oxon, OX14 4RN

CRC Press is an imprint of Taylor & Francis Group, LLC

© 2023 Klaas Sijtsma

ISBN: 978-1-032-18901-7 (hbk)
ISBN: 978-1-032-18374-9 (pbk)
ISBN: 978-1-003-25684-7 (ebk)

DOI: 10.1201/9781003256847

Typeset in Minion
by SPi Technologies India Pvt Ltd (Straive)

Contents

Acknowledgments

AFTER A LONG PERIOD of restraint, mainly fueled by having seen too much misery caused by research misconduct, I felt that after a couple of years and some mental distancing, I should write this book. Compared to other publications in scientific fraud and its softer but perhaps even more damaging cousin known as questionable research practices, this book is different in two ways. First, I emphasize researchers' poor mastery of methodology and statistical reasoning as one of the main causes of questionable research practices. Second, rather than proposing novel rules for using statistics, I promote the policy measures of preregistration of research plans and publication of data as well as seeking timely statistical advice. Situational influences such as career pressure and publication bias amplify the occurrence of questionable research practices, but the one-sided attention these effects receive in the relevant literature distracts from many researchers' poor mastery of methodology and statistical reasoning as a cause of questionable research practices.

To write the book, I have relied on my profession, which is applied statistics, and my experience as dean of a school suffering from the aftermath of a huge data fraud affair first exposed in 2011. Knowledge of statistics helped me to understand why statistics is such a slippery slope for many people. Having had to deal with fraud and questionable research practices as an administrator further sharpened my awareness of problems with using statistics. Insights from the best of what psychology has to offer further convinced me that, more than education, policy might help best to prevent fraud and support researchers adopting responsible conduct of research.

The book has three peculiarities. First, whereas misconduct and questionable research practices are everywhere in science across the world, my background as an applied statistician in a Dutch School of Social and Behavioral Sciences predisposed me toward examples and experiences from these substantive and geographical areas. I am confident these limitations will not pose a serious obstacle to what I have to say. Second, what you will read is an essay supported by insights from statistics, psychology, and administration, but not a formal derivation of a result from undisputed axioms. Thus, there is room for discussion and disagreement. Third, I discuss problems in research, suggest conditions for alleviating them, but do not solve them once and for all. This is a subject for continuous concern.

I am thankful to Marcel van Assen, Ilja van Beest, Lex Bouter, Jaap Denissen, Jules Ellis, Wilco Emons, Agneta Fischer, Patrick Groenen, Peter van der Heijden, Herbert Hoijtink, Ton Hol, Han van der Maas, Martijn Nolen, Michèle Nuijten, Hester Sijtsma, Tom Snijders, Marc van Veldhoven, and Jelte Wicherts, who provided valuable information that helped me write this book. Jaap Denissen, Wilco Emons, and three anonymous reviewers provided useful comments on previous book drafts. Wilco Emons programmed and discussed the simulation studies reported in Chapters 2, 4, and 5 and provided several figures. John Kimmel and Lara Spieker offered their valuable advice as publishers. Cindy Crumrine copyedited the text. The software we used for the simulation examples can be found in the Open Science Framework: https://osf.io/h2awx/. I am responsible for the flaws that remain in this book, and open for discussion.

Klaas Sijtsma, October 2022

CHAPTER 1

Why This Book?

THE REASON FOR WRITING THIS BOOK: A CASE STUDY

In September 2011, the breach of integrity of the man who was then the Dean of the School of Social and Behavioral Sciences at Tilburg University became headline news all over the world, and the university suddenly became well known in a way they would rather have missed. The Dean was Diederik Stapel, a famous Dutch social psychologist in his mid-40s, who stood in high regard among his colleagues in the Netherlands and elsewhere, and whom his students almost worshipped. By then, he had published more than 100 articles, mostly in high-profile scientific journals. Three young researchers from the Department of Social Psychology at Tilburg University, who failed to understand why their research results always came out with the usual imperfections whereas Stapel's were so invariably impeccable, decided they must get to the bottom of this remarkable contrast to find out what was going on. Realizing the sensitivity of their investigation, secretly and with the greatest caution, they collected the evidence across a couple of months, and then came up with compelling support for their hypothesis that something was terribly wrong with Stapel's research and reporting in scientific journals. When they felt they

DOI: 10.1201/9781003256847-1

knew enough, they took a brave step and managed to convince their supervising professor. He did not hesitate to confront Stapel directly, and based on the confrontation, informed the Rector Magnificus—comparable to the Provost at Anglo-Saxon universities—that same weekend. The Rector Magnificus immediately called up Stapel and invited him to his home that Saturday night to respond to the news brought to him. The exchange that followed gave the Rector Magnificus cause to inform confidentially the other members of the School's Management Team a week later about the suspected breach of integrity.

More than a year earlier, the previous Dean and his team had stepped down, and the University's Executive Board planned appointing Diederik Stapel as the new Dean of the School of Social and Behavioral Sciences. In a formal meeting, they asked the School's Department Heads for their opinion. I was the Head of the Department of Methodology and Statistics, so they also invited me. The Department Heads thought Stapel was a creative scientist and an enthusiastic propagator of his ideas and expected him to represent the School well but expressed concern about his limited experience as an administrator. They preferred the appointment of a Vice Dean who was an experienced administrator to compensate for Stapel's perceived lack of experience. Because I had been Head of the Department of Methodology and Statistics for a decade and had been active in other administrative and managerial positions, several of my colleagues looked at me. After some discussion, I agreed. A few days later, Stapel appointed me Vice Dean of Research. I started on 1 September 2010, together with Stapel, whom the Executive Board appointed Dean; the Vice Dean of Education, whom the previous Dean had appointed a few months earlier; and the newly appointed Managing Director. The four of us formed the School's new Management Team.

Precisely one year later, on 1 September 2011, together with the Managing Director, I was invited with extreme urgency to the Rector Magnificus' office, who then informed us that he had good reason to suspect Stapel of data fraud. The apparent fraud concerned a long period of approximately 15 years, including Stapel's affiliations with the University of Amsterdam in the 1990s and the University of Groningen from 2000 to 2006 before Tilburg University hired him in 2006. Moreover, the fraud to which Stapel already confessed implicated dozens of coauthors from many universities inside and outside the Netherlands, and worse, several former and present PhD students whose work might be seriously affected by their supervisor's misconduct. Because it was clear that Stapel would not return to his position as Dean and the University would probably terminate his contract, the Rector Magnificus asked me to consider stepping in as interim Dean and make the decision now so that I could start right away without anyone knowing for the next few days. I agreed.

The Rector Magnificus had already approached a retired professor of methodology and statistics and an active criminal law professor, both affiliated with Tilburg University, with the request that they become members of the committee that would investigate the alleged data fraud. Both had accepted. A few hours after I agreed to be interim Dean, the Rector Magnificus and I met with one of the nominated committee members. The University's regulations dictated the committee's Chair must be a professor from another academic institution. We very quickly agreed that a former President of the Royal Netherlands Academy of Arts and Sciences would be a convenient candidate and decided to ask Willem ("Pim") Levelt, a retired professor of experimental psycholinguistics affiliated with the Max Planck Institute for Psycholinguistics in Nijmegen. The Rector Magnificus made a phone call, and immediately realizing the seriousness of the situation and the stakes involved for all of psychology, Pim Levelt agreed without hesitation.

Looking back at the previous year in office serving under Stapel, I did not have any suspicion about his phony research activities. Before I became Vice Dean, except for a few short polite conversations, I hardly knew him but was aware of his reputation as a prominent scientist. I had very little idea what kind of research he did except he was a social psychologist, probably engaged in experimental research. As a statistician, I had little contact with colleagues who did experimental research. Experimenters in general had their own way of analyzing their data, often following a standard statistical analysis method known as analysis of variance, highly appropriate to analyze data from experimental designs, and they usually knew well how to do the analyses. Hence, they were not inclined to ask a statistician for help, and indeed no one ever asked my advice or that of my colleagues. As a Dean, Stapel was primarily focused on the big issues, such as realizing a new building for our School and discussing the School's future with the University's Executive Board. He left the tasks typical of their portfolio to the vice deans, hardly interfering. In my first year of what was to be a four-year term, my main tasks were creating a new research master curriculum replacing a program that would no longer receive state funding and preparing a self-study report for use in a national assessment of the School's psychology research program that took place every six years.

I remember one peculiar event, when during a weekly meeting, I noticed to my surprise that the other members of the Management Team had received a preprint of one of Stapel's papers that the famous journal *Science* had decided to publish. I had not received a copy. In the previous weeks, Stapel had been talking quite a bit about the paper's topic, which was about the influence a littered station hall had on people's tendency toward stereotyping and discrimination. Because I was not particularly interested in this kind of research and had many other things on my mind, I decided to ignore not receiving a copy. Only a few months later, after the Rector Magnificus informed me about Stapel's fraud, I learned that the *Science* article (Stapel & Lindenberg, 2011; retracted) was a complete hoax, made up from behind a desk, including the tables showing the alleged research results. Now I found myself in my new job as interim Dean writing an email to the Editor of *Science* informing him about the alleged fraud, thinking of the preprint I never received. Looking back, I wondered whether I would have noticed the paper's peculiarities had I received and read it. A fellow statistician I spoke about the article later said that when he read it, he wondered how Stapel collected the data at a complex research site such as a big and crowded railway station. However, the thought only crossed his mind, and he let it go. He did not notice anything unusual in the reported results, but probably did not look very closely. What this shows mostly is that it takes a giant step from wondering about one or two alleged inconsistencies to explicit suspicion.

The day the Executive Board informed me and informally appointed me interim Dean, together with the Managing Director, I also talked with the Vice Dean of Education, who was primarily angry, and the professor of social psychology, who informed the Rector Magnificus and who, understandably, was terribly upset. In the next few days, two issues were of particular importance to the University. The first issue was to get a formal confession from Stapel, so that the University could let him go. The second issue was to install the research committee as soon as possible and provide them with an official task. We—a small group of confidants chaired by the Rector Magnificus—considered both issues essential to pick a time to inform the university's community and the public. We wanted to bring the news quickly, because we felt we needed to inform coauthors, PhD students, and former PhD students and colleagues as soon as possible. Fortunately, a meeting of the Rector Magnificus and Stapel and the lawyers that represented both parties provided the formal grounds on which to terminate Stapel's contract. Installing the committee was easy once its members agreed to take part in what would undoubtedly become an intense and very difficult enterprise. Having taken both hurdles, we decided to go public.

On 7 September, six days after the Rector Magnificus informed me confidentially and little more than two weeks after Stapel's colleague informed the Rector Magnificus of the alleged fraud, he and I informed the University's community and the media.

That is, the Rector Magnificus was the spokesperson, and I stepped in when deemed necessary. At 10:00 AM, we informed the members of Stapel's Department of Social Psychology. While the Rector Magnificus spoke, I noticed how people reacted. I never saw a blow hit so hard. Colleagues looked bewildered when the Rector Magnificus broke the news, some panicked, and I could see others recover slowly and apparently begin to wonder what this would mean for the Department and for them. An hour later, we informed the Department Heads. Again, there were those bewildered gazes of initial disbelief and then the distinction between colleagues who sank away in contemplation and others who started moving around restlessly and talking. I sat opposite a professor whose eyes opened wider and wider and whose jaw sank lower and lower, until I thought he had reached the bottom of what was physically possible. He did not notice. Everybody was instructed to keep the news to herself until we had informed the whole School, which was what we did in a lecture hall, just after lunch. The hall was crowded, so people apparently had sensed that something big was going to happen. During the Rector Magnificus' speech, I could see people texting messages. After the meeting, colleagues stormed us and asked many questions. I remember our basic message was that Stapel would not return, and a committee immediately was going to investigate the alleged fraud and shed light on coauthors' and possibly others' roles. We emphasized that there was no suspicion whatsoever against anyone but Stapel, but that the investigation was necessary to ascertain the fraud, identify the publications that were fraudulent, and to clear the coauthors, whose reputations might otherwise remain tainted for the rest of their careers.

After we left the lecture hall, the University published a press release on their website. It took the media less than an hour to react. The first message we picked up from the Internet read "University kicks hoaxing professor in the street."[1] We could not help ourselves considering this good news. In the meantime, I tried to phone the deans of our sister schools from the University of Amsterdam and the University of Groningen, who were as deeply involved as we were, but probably had not been alarmed yet and needed to be updated about what was about to happen to them. Because I did not know yet who the Amsterdam Dean was, I called the director of research of the Psychology department, whom I knew, because he coordinated the national assessment of eight psychology research programs in which I was involved on behalf of our School and explained the situation to him. In Groningen, I failed to reach the Dean and decided to repeat the call as soon as possible.

By 5:00 PM, we found ourselves in a pressroom in one of the University buildings, where the Rector Magnificus would answer questions asked by a journalist from a national radio station broadcasting live. Other interviews with national newspapers followed quickly and, in the meantime, TV crews from a local station and the national 8 o'clock news entered the building and interviewed us simultaneously. Anyone worrying whether the 8 o'clock news still attracted viewers could rest assured. When I was on the train later that evening, travelling home after an unbelievably intense day, I received several emails from colleagues from other universities wishing me luck, and my oldest daughter called me to ask why I was on TV. When I came home by 10:00 PM, I picked up the phone and called the Dean in Groningen, whom I knew personally because like me he was a statistician and started updating him. I learned then what a head start of six days meant and heard him ask all the questions I had asked previously of the Rector Magnificus when he informed me. At midnight, I dropped into bed, and next morning could not remember ever hitting the pillow.

A BROADER PERSPECTIVE: INSTITUTIONAL FLAWS AND STATISTICS USE, THIS BOOK

In their seminal discussion of fraud in science, titled *Betrayers of the Truth. Fraud and Deceit in the Halls of Science*, William Broad and Nicholas Wade portray science as an enterprise committed to finding truth but geared toward a reward system that causes many of its practitioners to take shortcuts sometimes as bad as data fraud to achieve fame and fortune, or simply a living. The authors explain convincingly that the competitive nature of science directs appraisal of achievement toward counting numbers of published journal articles, no matter their content or relevance, rather than the quality of the work and the contribution it makes to the solution of real problems. As the number of researchers grows continuously, no one has time to keep track of the literature because too many papers are published in an ever-growing body of academic journals, many of which are below par, while everybody is too busy publishing their own work. Broad and Wade's message is so compelling and timely, if I had not known they published their book in 1982, I might have thought it was published yesterday (I read it in 2022!). What caught my attention in this and other books as well as numerous articles on the subject is the conviction that the cause of the malpractice lies in a variety of institutional flaws driving individuals' behavior when they do their research and publish their results. Although I acknowledge the point made, I do not completely agree with it. My stance is that not only institutional, hence external causes drive researchers' behavior but that other problems threaten to deteriorate the quality of published work as well. These problems rarely receive proper attention. This book fills that void.

Although they do not imply that every researcher suffers from external pressures to behave in certain ways, Broad and Wade and many other authors describe the scientific enterprise as a flawed institutional cage from which it is difficult to escape. The flaws originate from the clash of such antagonistic goals as science's noble quest for the truth and everyday practice of chasing personal credit and competition for scarce resources. They discuss and richly illustrate the reward mechanisms that lure researchers into wrongdoing, often without fully realizing what they do, such as the race to publish as much as they can to have tenure, continue financing their lab, hire more personnel, win their peers' admiration as well as prestigious prizes, and gain access to interesting people, events, and more resources. They also mention the hierarchy in research communities that easily assigns merit to the group's leader at the expense of the young researchers who did the work and made the discoveries. They even question the idealized goals of the scientific enterprise regularly presented in textbooks that they replace with the unruly practice of real research that strongly contradicts these ideals that nevertheless serve as beacons for imperfect research. This mismatch invites spicing up one's imperfect data and disappointing statistical results a little to obey the rules of perfection impossible to attain otherwise. And surprisingly, what goes wrong is often not intercepted and corrected but its gravity underestimated, ignored, or denied by those in charge.

Over the past years, the growing awareness of the presence of fraud in research and its less explicit but more damaging cousin known as questionable research practices has fostered a complete research area, almost exclusively emphasizing the deteriorating effect institutional flaws have on researchers. Much as I sympathize with the critical accounts of institutional causes of malpractice and the need to mitigate or remove these causes, being a statistician, I miss one cause of malpractice, especially when it is unintentional, which is researchers' lack of mastery of statistics. In my 40-year experience as an applied statistician, researchers' grave misunderstanding and incorrect use of statistics have never ceased to worry me, and I have seen little if any progress in all those years. But what do you expect? Someone can be an excellent medical or health researcher, psychologist, biologist, or biochemist, but this does not imply she is also an excellent statistician. Many researchers have at most attended a few introductory statistics courses in their bachelor's and perhaps one more course in their master's or graduate training, several have later attended one or more crash courses in a difficult subject such as structural equation modeling or multilevel analysis, and most can run statistical software, but does that mean they are able statisticians? I do not think so! Just as I am not an able physician, health scientist, psychologist, biologist, or biochemist, they are not able statisticians—exceptions noted, of course—but the odd expectation prevails that they can set up research, sample data, and analyze the sampled data as if they were real experts. This expectation is so common that it is difficult to even suggest it as an issue without running the risk of no longer being taken seriously.

However, among one another, many statisticians agree that researchers' poor mastery of statistics is one of the great problems if not the greatest problem in research involving data collection and data analysis. They do not refer to intentional fraud but rather to not mastering part of one's job, which is understanding and correctly applying statistics. I would expect this poor mastery is a real pain in the neck to researchers. Is this why many senior researchers make it a habit of leaving the statistical analysis of the data to a PhD student or even a master's student? Not that these younger colleagues are better statisticians, but at least it is off the seniors' desks. If statistics were simple arithmetic, this would not pose problems to academics, but statistics is much more than that, and it is difficult, while its chance mechanisms and results often run counter to intuition. I will argue that mastery of statistics requires not only immersing oneself in a serious statistics training program but also obtaining long-term experience that protects one from being constantly tricked by the counterintuitive results statistical analysis produces. The literature on what can go wrong in research focuses on institutional causes of malpractice, but in this book, I contend that researchers' poor mastery of statistics is perhaps an even bigger threat to the validity of research results. This position is not meant to be to the detriment of the researchers who are also able statisticians and researchers who honestly admit they cannot be good at everything and turn to a statistician for help. However, the purpose of the book is to raise discussion and awareness of the problem of having to do things at which you are not seriously trained and lack sufficient experience. An additional cause of malpractice I discuss is that probably spurred by the requirement to publish as much as one can, many researchers keep their data to themselves, making it impossible for others to check their work and, when necessary, correct it, a circumstance Broad and Wade also discuss. Thus, in the final chapter, Chapter 7, I will argue for open data policy and involving statistical expertise in research projects.

This book focuses on quantitative research, that is, research based on samples of quantitative data, because this is my expertise as an applied statistician and because most research is quantitative. I am not implying research based on qualitative data, such as case studies, interviews, archives, and documents that are not reduced to numbers cannot suffer from fraud and questionable research practices; it can, and examples are available as I illustrate briefly in Chapter 3. In Chapter 5, following other authors I argue that research using qualitative data should be preregistered just like research based on quantitative data, but this is a sideline of the general argument focusing on statistical analysis of numerical data. Other authors might take the opportunity to analyze the peculiarities of research using qualitative data, the threat of deceit in such research, and what to do to prevent malpractice as much as possible.

I discuss many examples of how a data analysis can trick you. Contrary to common practice using real data, I use artificial data I generated using a computer. This choice may surprise you, but I have a good reason for using artificial data. The point is that real data are excellent for demonstrating how to use a statistical model, for example, a linear regression model, and textbooks explaining this make ample use of real-data examples. However, real data are not well suited for showing how the statistical analysis of a sample of data can lead you astray. The reason is that real data do not enable you to assess the degree to which a sample deviates from the population from which you sampled your data and whether the model estimated from the data represents the true model for this population. Not knowing the true population distribution and the true model means you do not have a gold standard. You must realize that if you knew the population and the model, you would not have to do the research anyway! So, when I tell you that based on a real-dataset, a sample correlation of .51 looks very interesting but reflects a zero correlation in the population, you have no way of checking whether this is correct, and neither have I. This is different when I simply define (1) the population to be multivariate-normally distributed and (2) a linear regression model to be the true model, and then (3) use a computer to sample data consistent with the model from the population and (4) estimate the regression model from the data. Now you know that the sample deviates only by coincidence from the population, and can we discuss the influence of randomness on results that look interesting by comparing the sample results with the gold standard. I will keep everything as simple as possible in the hopes that the message—it is amazingly easy to be misled by data and statistics—comes through. I take this message up and explain it in Chapter 6 using insights from the best psychology has to offer.

In writing this book, in addition to relying on my experience as an applied statistician and some knowledge of the relevant literature, I also relied on my experience as a dean of a school plagued by the aftermath of the unmasking of a colleague who was well respected until he was found to be a data fraudster. I will discuss what cases like this do with an academic environment in Chapter 3 and focus on the methodological and statistical issues involved in different kinds of data cheating in Chapters 3 and 4. Prior to that, I discuss in Chapter 2 the concepts needed for a better understanding of Chapters 3 and 4 and later chapters, as well as a couple of examples based on artificial data that show the deception statistical results based on samples can provide. Readers familiar with the concepts and the deception brought about by data analysis can choose to skip Chapter 2, but I recommend reading it as well before moving on. Statistical analyses and explanations of statistical principles in later chapters have sometimes been moved to the appendices but I trust that motivated readers are prepared to work their way through the intricacies of statistical reasoning, if necessary. A summary of the next six chapters is the following.

Chapter 2 discusses the book's focus on inexperienced or misinformed use of statistics as an important source of questionable research practices. Chapter 3 follows up on the Stapel fraud case and discusses how this case increased researchers' awareness and readiness to take precautions against questionable research practices. In Chapter 4, I discuss the falsification and fabrication of data in a simple way, showing that even though such data may look credible, they contain features that help to expose them. Chapter 5 discusses the superiority of confirmatory to exploratory research, but notices that most research is exploratory, fulfilling a valuable role in science under certain conditions. In Chapter 6, I argue that most questionable research practices happen because many researchers using statistics underestimate its difficulty and are misled by statistics' counterintuitive results; in short, they lack the experience to practice the profession of statistical data analysis well. Chapter 7 discusses several measures to reduce the occurrence of questionable research practices.

I hope a broad readership consisting of students, researchers, statisticians and interested laypeople will learn from this book or use it for engaging in further discussion. Chapters 2, 4, 6, and 7 have technical appendices discussing background knowledge for a better understanding of some aspects of the main texts, and materials backing up the simulation of artificial data. The appendices were added for the readers' convenience but can be skipped by those already well versed in basic statistical theory and application. Each chapter ends with three take-home messages intended to summarize key information and raise discussion.

NOTE

1 https://www.geenstijl.nl/2692841/tilburg_university_gooit_hoaxp/

CHAPTER 2

Fraud and Questionable Research Practices

*F*RAUD AND *ERRATICISM* ARE two kinds of inappropriate behavior that threaten the credibility of scientific results. Fraud refers to careful and rational decision-making with intention to deceive colleagues and readers of scientific reports. Erraticism refers to behavior being guided by patterns of misunderstanding or misinformation without intention to mislead but nevertheless resulting in flawed research results. The main distinction thus is whether one's intentions are deliberate or non-deliberate. In layman's language, we are dealing here with the difference between swindle and clumsiness or, phrased kindlier, unfortunate action not resulting from intention of wrongdoing but often from the best of intentions but with the wrong effects. We should condemn the swindler and help the unfortunate fellow researcher. This is the approach followed in this monograph.

The meaning of the word *fraud* will not lead to much debate, but erraticism may need some explanation. I chose the word

DOI: 10.1201/9781003256847-2   **15** *erraticism* to avoid researchers feeling depreciated or belittled when using terms such as ignorance, incompetence, clumsiness, or sloppiness. An erraticism is rather an odd or peculiar habit,[1] the action or tendency to be erratic,[2] in short, an act that is incorrect in one way or another and that repeats on several occasions. Key to erraticism is habit or tendency, rendering it a structural problem. I distinguish erraticism from *accident*, which is an unforeseen and unplanned event or circumstance, without intention or necessity,[3] happening completely by chance.[4] Hence, accidents are incidental problems. When an accident happens in scientific research and when it is recognized, one should correct it, but in general, an accident is much less of a problem to the validity of research results than an erraticism. In this monograph, I focus on erraticism.

More than fraud, erraticism is my focus here as well as on the error it produces in research and the reporting of research results, for two reasons. First, even though fraud can be terribly damaging to science's credibility, fraudsters often operate alone, not infecting a whole research area, whereas erraticism often is widespread among researchers' communities in the form of incorrect research habits and may slowly undermine the research area. Second, like all people intentionally doing something illegal for benefit, the fraudster will do it anyway given opportunity and little chance of being caught and secretively lead colleagues astray, whereas the researcher making mistakes without bad intention in principle is open to help from colleagues providing methodological and statistical expertise.

METHODOLOGICAL AND STATISTICAL MISSTEPS

Fraud: Fabrication, Falsification, and Plagiarism

Fraud in science refers to misconduct aimed at deliberately presenting a desirable finding one never obtained, because one's data do not support this finding. Examples are making up one's data—fabrication—to obtain planned and desirable results and changing one's data—falsification—to obtain better results. Falsification also includes changing numbers in tables and curves in graphs, and particular details in brain scans, X-rays, and photos to suggest a result not present in the original source. At the extreme, one makes up the entire research and publishes an article about a research project that never happened (e.g., Stapel & Lindenberg, 2011; retracted) and suggests a finding that the author expects will bring her fame and resources or at least a good reputation with her colleagues and funding organizations. Fraud also includes presenting texts and results one copied from other authors without mentioning the source, known as plagiarism. Plagiarism is intellectual theft, but often concerns valid texts or research results and in that sense does not do damage to the body of scientific knowledge. However, plagiarism undermines the trust in colleagues and in science.

I assume researchers are aware they engage in fabrication, falsification, or plagiarism. Ethical orientations researchers may adopt when considering how to adjust in a research community (Johnson & Ecklund, 2016) acknowledge the possibility of misconduct. The consequentialist ethical orientation is based on perceived practical consequences of certain actions that may or may not discourage the researcher engaging in activities such as misconduct, but the deontological ethical orientation builds on autonomous moral reasoning and follows certain principles the researcher considers serving the general good. In the context of business research, Fink, Gartner, Harms, and Hatak (2023) found that compared to deontology-oriented researchers, consequentialist-oriented researchers were more susceptible to misconduct the stronger the competition for resources. The virtue ethical orientation adopts practical wisdom, finding the right middle path between several options, avoiding extreme actions. My guess is that the virtue orientation nowadays can count on little sympathy when it comes to research practices not according to the book albeit not as serious as downright fraud, but De Vries, Anderson, and Martinson (2006) provide an interesting account of researchers trying to cope with this gray area of research glitches.

The topic of ethical orientation may provide more insight in researchers' motives, but I will let it rest, rather focusing on intention and problems showing it.

In real life, one does not have direct access to somebody else's motives, and therefore committees looking into cases of alleged fraud must prove suspicion of fraud beyond reasonable doubt. Intention is evident when fabricating part or the whole of a dataset without having collected data with real respondents or making up an article about a study that never took place. Showing beyond reasonable doubt that leaving an experimental condition out of a graph qualifies as fraud is more difficult. The researcher may claim that an incorrect measurement procedure contaminated the measurements in the deleted condition, considered the remaining results worthwhile publishing, but then forgot to include a footnote explaining the omission of the contaminated condition. Whereas such an omission is undesirable, based on the available evidence, an integrity committee looking into it may conclude it concerned an *accident*. If, however, the committee finds a pattern of omissions in several of the researcher's articles, sometimes to the benefit and sometimes to the detriment of the researcher, it may conclude that the repetition of these instances renders *erraticism* more likely. When the pattern always favors outcomes deemed desirable but unsupported by the data, the verdict of *fraud* comes in sight, but other explanations pointing in the direction of erraticism are still possible. Complicating circumstances are that the researcher has a habit of following a flawed procedure (e.g., *p*-hacking; e.g., Wicherts, Veldkamp, Augusteijn, Bakker, Van Aert et al., 2016), uses an obsolete and biased statistical procedure (e.g., for handling missing data; see Van Ginkel, Sijtsma, Van der Ark, & Vermunt, 2010), or lets inexperienced research assistants prepare the data by removing anomalies (e.g., scores out of range, missing data, outliers). Fraud is only easy to establish in some extreme cases, but in other cases, more difficult or impossible, thus slipping into a gray area where right and wrong are difficult to distinguish from each other (Johnson & Ecklund, 2016).

RCR ---------- Accidents, Erraticism (QRPs) ---------- FFP

FIGURE 2.1 The continuum of research running from responsible conduct of research (RCR) via accidental errors and erraticism (questionable research practices; QRPs) to fabrication, falsification, and plagiarism (FFP).

Unveiling fraud and making it public has the effect of a raging fire, bringing huge damage to science's credibility and reputation, and sometimes the institution where the fraud took place. In the interest of truth finding, which is what science stands for, scientific fraud must be fought and banned, but when it happens, made public, investigated to the core, and corrected. Correction may involve the retraction of articles from scientific journals, including a warning to readers that the journal retracted the article and the reason why they did this. In the literature (e.g., Steneck, 2006), the variety of fraudulent activities I have mentioned go under the name of Fabrication, Falsification, and Plagiarism (FFP). Its opposite, which is research according to the textbook, is Responsible Conduct of Research (RCR). Between the two opposites is a vast gray area where real-world research takes place with all its shortcomings, such as accidents and erraticism. A common name for erraticism is questionable research practices (QRPs). Figure 2.1 shows the continuum running from RCR via QRPs to FFP.

Accidents

Accidents happen frequently and are part of human nature. Our brain is a kind of computer, to use the metaphor, but one that is extremely slow, resistant to dull and repetitive tasks, and liable to make errors; that is, everything an algorithm executed by a computer is not. The great feature that makes our brain stand out is its capacity to learn from experience, resulting in extreme flexibility. This is what artificial intelligence tries to copy. Making errors is important here. Specifically, I am concerned with errors that can always happen and that one cannot prevent from happening, simply because research is complex, and we are in possession of an error-prone brain rendering us fallible creatures. An example of an accident is entering incorrect specifications into a software program that go unnoticed and do not prevent the software from producing output that looks credible, so that the researcher may believe it and does not become suspicious. Another example is accidentally copying results incorrectly from the correct output in a table that appears reasonable to a degree that reviewers and readers will not notice anything peculiar. Such errors can have serious consequences but seem unavoidable. Mellenbergh (2019) gives an overview of errors that are common and persistent and discusses methods for preventing or correcting the errors.

The examples also suggest that the researcher can never be careful enough and should never stop checking her work for errors, but there are practical limits to being careful. Not working alone but in a team where the team members are coauthors being responsible for what they publish in their names provides additional pairs of eyes that help to produce work that is as impeccable as possible. However, accidents will happen unavoidably. If detected, the authors should publish a correction when the error seriously affected the research outcomes. Researchers who are simply sloppy by design of character, paying too little attention to what they do, have little chance of a successful career in research and will probably leave research after a while. I will leave accidents, due to their incidental nature, at rest in the remainder of this book.

Erraticism: Questionable Research Practices

I am particularly interested in the errors originating from habit that tend to repeat across different studies by the same researcher and that larger groups of researchers adopt, believing they are doing the right thing. Because errors originating from habit are structural and repetitive, they drive research results and in the worst case a whole research area in the wrong direction. This can have the effect that for a long time a particular result passes for standard in a research area until somebody finds out it is the product of incorrect reasoning, methodological flaws, or faulty statistical data analysis. Undetected structural errors initiate follow-up research by others who take the flawed result seriously and use it as a starting point for their own research. Because it may take a while before the erroneous result becomes apparent, much time and resources of several researchers are wasted (Ioannidis, 2005; Steneck, 2006; however, see Ulrich & Miller, 2020). Obviously, we must avoid structural errors as much possible.

Numerous causes lie at the origin of structural errors collected in the category of QRPs. These causes often but not exclusively concern external circumstances that tempt researchers to engage in undesirable behavior. Examples are a university's pressure on researchers to publish as much as possible, leading to haste and inaccurate work; the policy of journals to publish only interesting results, ignoring the work that does not find such results, thus presenting a biased picture; and the temptation of access to research resources, fame, and status, leading a researcher to take shortcuts that lead to inaccurate results. The effects of different external causes of QRPs are diverse. A hasty researcher may consistently use samples too small to produce reliable results and pick out the accidental fruits, thus allowing publication of unreliable results. Another researcher may decide not to submit a concept article for publication in a journal because it contains an austere albeit correct result the author anticipates journals will not publish, also known as the file drawer problem (Rosenthal, 1979). Finally, a successful researcher having become reckless and insensitive to colleagues' critical comments may trust her expectations concerning research outcomes too much and underestimate the complexity of a data analysis, thus publishing incorrect results. For further discussion, see Anderson, Ronning, De Vries, and Martinson (2007), Fanelli (2009), Haven, Bouter, Smulders, and Tijdink (2019), Van de Schoot, Winter, Griffioen, Grimmelikhuijsen, Arts et al. (2021), Gopalakrishna, ter Riet, Vink, Stoop, Wicherts et al. (2022), and Miedema (2022).

An interesting question rarely discussed in the literature is why certain external or situational influences on research cause QRPs. Why would a researcher who uses statistics well without such external influences resort to QRPs when faced with such influences? Do performance pressure, the file drawer problem, complexity underestimation, and other situational causes have the same deteriorating effect on researchers mastering statistics well as on their colleagues who are not statistics experts? Is the degree in which researchers master statistical thinking a factor in the occurrence of QRPs? Several authors (e.g., Campbell, 1974, 2002; Gardenier & Resnik, 2002; Hand, 2014; Kahneman, 2011; Salsburg, 2017) have pointed out that researchers and even trained statisticians have trouble understanding the counter-intuitive results probability and statistics produce. They explain that this easily leads to errors in data analysis using statistical methods and procedures. My *proposition* in this book is that inexpert statistics-use is an important and often overlooked cause of QRPs, the occurrence of which may be bolstered further by performance pressure, publication bias, complexity underestimation, and other situational influences. I will not claim that inexpert statistics-use is the only cause of QRPs, but I contend that, for example, performance pressure is not likely to cause QRPs when the researcher masters statistics well. Why would external causes of QRPs force a researcher well versed in statistics to make errors in applying statistics she would not make without these causes? If she would, I would suspect intentions of deliberate fraud simply because she knows what she is doing.

I focus on inexpert statistics use and discuss three examples of statistical analysis, each producing invalid results. One example concerns a typical QRP, and the other two examples show that use of statistical procedures that are formally correct or widely accepted, respectively, may require profound statistical expertise for their correct application. The first example concerns leaving out data to produce a significant result, an instance of p-hacking. The second example concerns a multiple-regression optimization procedure that requires experience to produce meaningful results. The third example concerns the processing of missing data using obsolete and inferior methods, thus producing biased results. Together the examples show how easily the use of statistics produces incorrect or at least questionable results. For statisticians, the examples refer to well-known deficiencies, and a literature search will produce several useful references. Appendix 2.1 discusses basic methodological concepts.

Example 1: Outliers

The example shows that using statistics without external support by either a scientific theory—here, not applicable—or a client— here, an imaginary City Council—easily gets out of control. I start with this example as a first attempt to argue that even a simple statistical data analysis is difficult and invites error all too easily. Then, I move on to a case where observations are removed from the data because the imaginary researcher incorrectly believes this is justified to help amplify a difference between two groups, which she believes is real and must be revealed. This example illustrates a QRP.

Suppose the City Council wants to know the average household income in the municipality, broken down by neighborhood, to determine which neighborhoods need social and financial support the most. Your neighborhood, which I call B, is small, inhabited by many typical middle-class families who are well-educated, most of them in their thirties and forties, some older, some living alone, but many organized in households with children living at home. Table 2.1 shows the annual incomes of a random sample of 25 households from your neighborhood (B, right-hand panel). Visual inspection shows incomes range from $23,180 (# 38) to $66,020 (# 50), with a mean of $47,002.

I assume that your neighborhood accommodates a household not included in the sample and having an exceptionally high income due to participation in a successful family business. Let us say that their income is ten times the highest income in the

TABLE 2.1 Annual Household Incomes in Neighborhoods A and B

Neighborhood A

1	34.65	11	43.98	21	70.97	
2	51.03	12	54.20	22	41.95	
3	47.81	13	51.57	23	47.57	
4	45.17	14	50.08	24	56.32	
5	47.23	15	51.44	25	57.18	
6	39.07	16	43.17			
7	63.16	17	51.12			
8	49.11	18	55.09			
9	51.42	19	70.79			
10	46.57	20	69.18			

Neighborhood B

26	36.03	36	57.20	46	47.38	
27	46.43	37	34.91	47	54.59	
28	44.95	38	23.18	48	56.66	
29	43.97	39	29.68	49	60.02	
30	48.22	40	37.86	50	66.02	
31	30.70	41	43.18			
32	53.60	42	64.77			
33	41.19	43	41.27			
34	48.85	44	48.71			
35	47.56	45	60.11			

Note: Multiply entries by 1,000.

sample ($66,020, # 50), resulting in a $660,200 income. Suppose the $660,200 income rather than the $66,020 (# 50) would have been sampled, then the sample mean had been $89,787, nearly doubling the original mean of $47,002. This inflated sample mean is much larger than each of the other 24 incomes. Obviously, sampling the $660,200 income produces a mean not representative of the other incomes and is possibly disadvantageous with respect to municipal policy for the whole neighborhood.

Should the researcher exclude the one top income from the investigation? The example is quite extreme and the answer in this case obvious, but it would help the researcher a lot when, prior to collecting the data the City Council defines what they consider a valid mean neighborhood household-income suitable for making a policy decision. Suppose the City Council is of the opinion that one or a few extremely high incomes must not determine conclusions about the large majority's income situation, without specifying what they consider extreme. Then the researcher may decide that the x percent highest incomes are excluded from the subset of cases eligible for sampling. This subset is known as the sampling frame. The problem is choosing x, which involves an arbitrary choice, and the researcher may report results for several values so that the choice problem becomes obvious. Another approach would be not to exclude any incomes but use a statistical solution such as reporting the median sample income. To obtain the median, order the sampled incomes from lowest to highest and take the one in the middle, the median, as the typical neighborhood income. In the example, the median is $47,560 (# 35) with and without the highest income. Whatever the approach, the researcher must report results obtained under the various scenarios she considered. Preregistration of the planned research is another possibility; see Chapter 5.

Outliers are often more difficult to recognize than in the income example and easy to misrepresent, as I will also illustrate in Chapter 5 in the context of an experiment. So far, I have not asked what an outlier is, but you will have some intuition on which

I have relied thus far. According to a standard work on outliers (Barnett & Lewis, 1994, p. 7) an outlier is "an observation (or subset of observations) which appears to be inconsistent with the remainder of that set of data." Ignoring the apparent obviousness of the definition as well as the complexity involved in an analysis of outliers, I settle for the conclusion that to assess whether an observation is an outlier, one needs to study it in relation to the other data. The issue is whether an observation seems to belong to the population of observations, not whether it increases or decreases the differences between two group means or another result of interest to the researcher. Thus, you should study the occurrence of outliers prior to the data analysis directed at answering the research question and, based on available knowledge about sources that produce outliers, decide which observations are outliers and should be handled as such in the data analysis. You must not first have a look at the outcome of the statistical analysis and, when displeasing, identify observations that might embellish the outcome when left out of the analysis.

To illustrate the problem of *ad hoc* leaving out observations based on knowledge of the outcome of statistical analysis, I constructed an example using artificial data (Table 2.1) that I simulated from hypothesized statistical models. I used simulated data to have an example I can completely control to make my point. The simulated data do not bear any resemblance to real data unless by coincidence, and serve only educational purposes. I will illustrate only what can go wrong if one deletes data from a dataset to see what happens to the outcome of interest. Readers who want to know how to analyze their data correctly and recognize potential outliers can consult the literature on methodology and statistics. The problem is not that this information is unavailable—it is available, in vast amounts—but that researchers often are unaware of it, ignore it, or once knew but partly forgot it, or simply misunderstood and incorrectly applied it.

Suppose you expect that based on neighborhood characteristics, the mean annual income in Neighborhood A is higher than in Neighborhood B and want to find out whether this is true. Table 2.1 shows the incomes in two random samples of size 25, each drawn from the two neighborhoods. Indeed, the mean-income difference is in the expected direction (Table 2.2, first line). Because two samples do not provide conclusive evidence whether there is a real income difference in the expected direction, you perform a one-sided independent-samples Student's t-test (with Welch correction[5]). Researchers consider a sample difference D or larger having probability p smaller than .05, given the *null hypothesis* that the two complete neighborhoods have the same mean income, as evidence of a *real* income difference. You may notice that sample research cannot prove a hypothesis, implying that evidence only means support (Abelson, 1995). In Chapter 4, I discuss the logic of null-hypothesis testing in more detail. Here, I follow a simplified approach often found in research.

Table 2.2 (first line) shows that for the sample difference, $p = .0579$. According to the logic of statistical testing, you cannot conclude that Neighborhood A has the highest income. Because this value is so close to .05, you may wonder whether leaving out the lowest income from the highest-income sample or the highest income from the lowest-income sample (or both) pushes p below .05, a desirable outcome given your expectation (e.g., Head, Holman, Lanfear, Kahn, & Jennions, 2015; Masicampo & Lalande, 2012). After all, should one observation determine being able to draw a conclusion? Although this is a sensible question and it is easy to understand why you would ask it, blindly following this path implies a QRP. The crucial insight is this: Given that you have seen the outcome of the comparison, changing the data composition next by leaving out observations is like placing bets on a horse after the race is over, a typical case of foul play. For the sake of argument, I will assume that the imaginary researcher is motivated to report a difference to her employer, the City Council, and does this by leaving out the lowest income from Group A (second line), which produces $p = .0335$, a value small enough to reject the null hypothesis of equal mean incomes. She reports this result.

TABLE 2.2 Sample Means, Mean Difference, and Statistical Test Results, for Full Data and Data with Observations Left Out

Data	Mean A	Mean B	D	t	df	p
Full dataset	51.5932	47.0016	4.5916	1.6026	46.692	0.0579
Without lowest in A	52.2992	47.0016	5.2976	1.8772	45.494	0.0335
Without highest in B	51.5932	46.2092	5.3840	1.9104	45.816	0.0312
Leaving out both	52.2992	46.2092	6.09	2.1949	44.623	0.0167

Note: Income means and differences must be multiplied by 1,000.

She finds similar results when leaving out the highest income from Neighborhood B (third line) and obviously when leaving out both incomes (fourth line). You know now she should not report results as if she found them without any data manipulation.

Here is some background information on the data simulation, which I saved on purpose till last. The annual income in the neighborhoods has precisely the same distribution, which is normal with mean equal to $50,000 and standard deviation equal to $10,000. This means that 95.45% of the incomes lie between $30,000 and $70,000, whereas 99.7% lie between $20,000 and $80,000. I drew two samples from this distribution and called the sample with the highest mean Neighborhood A. The two "outliers" the researcher left out were not even real outliers, but the message that should stick is that it does not make sense to leave data out once the statistical outcome is known. Another message that will return repeatedly in this book is that it is very easy—and understandable—for researchers to make errors in statistical data analysis.

Example 2: Multiple Regression

Next, I discuss the popular method of multiple regression analysis (Fox, 1997; Pituch & Stevens, 2016) to study the relation between a set of variables known as predictors or independent variables on the one hand, and one variable known as the criterion or dependent variable on the other hand. Next, I demonstrate how a "mindless" algorithm produces flawed results with the otherwise correct multiple regression method. Researchers understandably trust the algorithm and the software and cannot be blamed for results that can be flawed, a consequence of sampling error that easily deceives them. The statistical definition of the multiple regression model is in Appendix 2.2. Here and elsewhere in the book I try to bypass statistical notation and explanation of methods and procedures to keep things simple. Readers interested in statistical background may consult the appendices I provide, others may ignore the appendices and move on.

In the example I will discuss, I consider ten predictor variables that are used to obtain the best prediction of a criterion variable. A predictor variable is more important for predicting the criterion variable the higher it correlates with the criterion variable and the lower it correlates with the other nine predictor variables. Based on these correlations, each predictor variable receives a weight estimated from the sample data. The scores of the sample units on the predictor are multiplied by this weight. In words (and simplifying a little), the multiple regression model looks like

$$\text{Weight}_1 \times \text{Predictor}_1 + \text{Weight}_2 \times \text{Predictor}_2 + \ldots$$
$$+ \text{Weight}_{10} \times \text{Predictor}_{10} = \text{Criterion}.$$

A researcher might study the relation between student proficiency and environmental influences on the one hand and success in college on the other hand. The predictor variables could be High School Performance (Predictor_1), SAT Score (Predictor_2), Education Parents (Predictor_3), ..., Number of Siblings (Predictor_{10}), and the criterion variable could be Success in College expressed as examination results at the end of the first year. The weights are estimated from a sample that provide scores for each predictor variable and the criterion variable for each student.

It is important to realize that the prediction of the criterion variable is never perfect, a situation typical of the human sciences.[6] The degree to which the prediction succeeds can be expressed as the correlation between the real examination results in the sample and the examination results as the ten predictor variables forecast them. This is the multiple correlation, denoted R. We use the squared multiple correlation, R^2, that expresses the proportion the estimated multiple regression model explains of the variance of the real criterion scores the sampled students provided. One can use statistical testing to establish whether sample weights provide evidence that the corresponding population weights are different from zero, meaning they provide a statistically significant contribution to criterion prediction.

Confirmatory research derives a multiple regression model from a theory about a phenomenon of interest and then collects data to test the model's tenability against the data. One may even have an *a priori* idea about the relative importance of the predictor weights and check in the data whether such detailed expectations are correct. For example, the model with ten predictor variables for college success may represent a theory in which these ten predictors are fixed based on several preliminary studies, and the model is expected to explain a high percentage of variance of college success. Here, I focus on *exploratory* research, dominant in many research areas, which considers the population model unknown and estimates a model from the available sample data. In this case, the ten predictors are simply available, not the result of painstaking preliminary research supporting theory construction, and the final preferred model may contain only a predictor subset that was determined by statistical criteria. Considerations such as efficiency, including only the predictors in the model that the data, not the theory, indicate contribute most to predicting the criterion, are prominent in much research. Confirmatory and exploratory research are central in Chapter 5.

In my example, I have ten predictors available. This is not the example concerning college success; the number of ten predictor variables is simply convenient, that is, large enough, to make my point. A search procedure called backward stepwise regression starts with all ten predictors and estimates the model. If it exists, the procedure eliminates from the model the predictor that contributes least and non-significantly to predicting the criterion. When left out, R^2 does not decrease significantly, meaning the smaller nine-predictor model has not lost predictive power and replaces the ten-predictor model. Then, the procedure continues with the smaller model; otherwise, the model with ten predictors is final. If dropped, the nine-predictor model is estimated, and the procedure looks for the predictor that contributes least and tests whether it contributes significantly to predicting the criterion. If it does not contribute significantly, it is dropped from the model and the procedure continues with the eight-predictor model; otherwise, the nine-predictor model is final. The procedure continues until a model remains without predictors that do not contribute to predicting the criterion.

Before I examine results produced by the data-oriented backward stepwise regression method, two words of caution are in order. First, if I simply include all predictors available in the model, this produces the largest R^2, but this approach ignores efficiency considerations by possibly including predictors that do not contribute significantly to explaining the criterion. Second, because the human sciences often lack well-developed theories, data analysis frequently is exploratory, inviting increased risk of ending up with results relying too much on sampling error and not resistant to replication. I use an artificial example to illustrate how this works with backward stepwise regression.

I assume the ten predictors correlate .1 among one another and show the same positive correlation with the criterion, equal to .25. This means that each predictor provides an equally sized unique contribution to predicting the criterion. Thus, all predictors are equally important, a rare situation in practice but instructive for the point I wish to make. The predictors are represented by numbers, 1, ..., 10. All predictors and the criterion Y have mean 0 and variances equal to 1. The first line of Table 2.3 shows the standardized (implying they are comparable) regression weights, which are all equal. This is the population model.

For each imaginary respondent, I sampled data lines with eleven scores from a multivariate normal distribution and repeated this for sample sizes N = 50, 100, 200, 400, 1,000. Population correlations are all .1, and Table 2.4 shows the randomly differing sample correlation structure for N = 50 with correlations varying from −.22 to. 40 (both values printed in bold to facilitate identification). Because the sample correlations differ

TABLE 2.3 Population Predictor Weights for the Full Model (First Line), Sample Predictor Weights for Models Selected Using Backward Stepwise Regression for Different Sample Sizes N (Blanks: Predictor Not Included in Model), Multiple Correlations (R^2, R^2_{repl})

N	1	2	3	4	5	6	7	8	9	10	R^2	R^2_{repl}
	0.13	0.13	0.13	0.13	0.13	0.13	0.13	0.13	0.13	0.13		
50	0.33		0.29		0.32			0.22	0.22		.32	.16
100	0.29	0.29				0.25		0.23	0.23		.36	.23
200		0.20	0.16	0.19		0.17				0.14	.38	.35
400	0.13		0.19	0.16	0.11	0.12	0.12	0.16	0.15	0.19	.35	.30
1000	0.13	0.13	0.19	0.09	0.11	0.13	0.15	0.10	0.13	0.11	.30	.30

from the population correlations that generated them, the point of departure for the explorative analysis differs from the population model. Thus, purely by chance, I might find a model different from the population model. For each sample size, I considered the final models resulting from backward stepwise regression. Their estimated regression weights are in Table 2.3. For $N = 50$, one sees that only three predictors were selected and that their regression weights are a multiple of the population weights. Random sampling and nothing else caused this highly deviant result, hardly providing evidence of the population model. Larger sample sizes produce models that deviate less, but only $N = 1,000$ produces a model almost consistent with the population model. That is, backward stepwise regression selected all predictors, but their weights still show variation.

I computed multiple correlations between criterion scores sampled from the multivariate normal distribution and predicted criterion scores using the final models resulting from the stepwise method. They varied little. Next, I drew new samples representing replications, and for each sample size, I multiplied the regression weights of the model selected in the first round with the corresponding predictor scores sampled in the second round to estimate criterion scores sampled in the second round. Table 2.3 shows R^2_{repl}, the multiple correlation of the estimated criterion in the replicated sample and the sampled criterion score, running from much smaller to almost equal to R^2 as sample size increases.

What can we learn from this example? First, one could argue that especially in the absence of a theoretical model, selecting the best predictors is useful for predicting the criterion as well as possible. However, a counter-argument is that an explorative model will not stand ground with new data unless the sample is large (e.g., $N = 1,000$), and this is what the replicated data show. For smaller samples, information is simply uninformative about the population model. Second, one may argue that the selected model is at least informative about the sample. True, but this would mean giving up the ambition to generalize a result beyond

TABLE 2.4 Sample ($N = 50$) Correlations in Upper Triangle (Lower Triangle Holds the Same Values, Not Shown; Smallest and Largest Values Printed Boldface to Facilitate Identification)

	1	2	3	4	5	6	7	8	9	10
Y	.37	.17	.33	.33	.31	.16	.23	.14	.11	.11
1		-.17	.15	.19	-.00	-.02	-.08	.16	**-.22**	-.03
2			-.01	.02	.17	-.05	.09	-.21	.04	-.22
3				.18	-.05	.12	.11	.13	.26	.14
4					.23	-.02	.11	.05	.07	.19
5						.21	.09	.03	.06	.05
6							.28	**.40**	-.10	.10
7								.00	.13	-.03
8									-.01	-.01
9										-.01

the sample, a situation rarely realistic. I do not suggest that exploration is useless, only that it should provide a start of a more ambitious project aimed at developing the real model, but no more than a start.

I predict that some experienced data analysts will tell you that I should have used backward stepwise regression differently, used another variation of the method, or used a completely different method. However, I repeat that the point I wish to make is that the combination of multiple variables, small to modest sample sizes, and automated model selection readily leads you astray, especially when a guiding theory is absent.

Example 3: Missing Data

I discuss the use of two popular methods for dealing with the problem of missing data. This problem happens when some respondents fail to provide answers to one or more questions from a larger set. The methods are listwise deletion and available-case analysis, both very simple, sometimes effective but often sub-optimal (Schafer & Graham, 2002). I discuss two issues using these popular methods. First, researchers seem unaware of the methods' statistical inferiority to other available methods. Second, if they use the methods (or better alternatives), they often fail to check the conditions for the methods' responsible use in the data. This failure may cause flawed results (Van Ginkel et al., 2010).

Figure 2.2 shows five questions from a longer questionnaire typically used in the human sciences, collecting information from respondents answering the questions from home on their computer. The questions target consumer behavior. The researcher uses the answers to study the relationships between the questions, thus learning about consumer habits. This is survey research; also, see Chapter 5. Typically, some respondents fail to provide answers to some of the questions, usually for different reasons. For example, some respondents did not understand a particular question ("What percentage of your income did you spend on consumer

Q1—What percentage of your income did you spend on consumer goods last year?	------------ %	No more than 15% ☐ 16—25% ☐ 26—35% ☐ 36—45% ☐ More than 45% ☐
Q2—How much money did you spend last year on food?	$ ------------	Less than $2000 ☐ $2000—$3000 ☐ $3001—$4000 ☐ $4001—$5000 ☐ More than $5000 ☐
Q3—What was your gross income last year?	$ ------------	Under $15,000 ☐ $15,000—$25,000 ☐ $25,001—$35,000 ☐ $35,001—$45,000 ☐ $45,001—$55,000 ☐ More than $55,000 ☐
Q4—Did you ever take something from a shop without paying?	------------	☐ Yes ☐ No
Q5—What is your highest level of education?	------------	☐ Elementary School ☐ Middle School ☐ High School ☐ Vocational ☐ University ☐ None

FIGURE 2.2   Survey questions (Q1 = question 1, etc.), open-responses, and constructed-responses.

goods last year?" The problem is that words are not well defined: percentage, income, consumer goods); understood the question but did not know the answer ("How much money did you spend last year on food?"); or found the question impertinent ("Did you ever take something from a shop without paying?"). Trivially, some respondents may have skipped the question at first intending to get back later to it and then forgot ("What is your highest level of education?").

The respondent is asked to fill out (Figure 2.2, second column) a percentage, a sum of money, an income, yes/no or maybe, forgot, don't know, and so forth, and an education type. This open-response mode may invite inaccuracies when respondents are not sure and guess, or nonsensical answers, such as "129 years" when asked their age. The third column shows a constructed-response mode, in which the respondent choses from several given answers the one that applies to them the best by clicking on the square boxes. Constructed-response questions avoid odd answers. They also allow easy transformation to scores needed for statistical analyses.

Figure 2.3 shows a fictional numerical dataset (for the moment, ignore the numbers in bold between columns) with scores assigned to the constructed-response mode (score ranges provided in figure caption). Blanks represent missing values. Incomplete datasets are problematic. I consider the first six data lines and Q1—Q5 as the data for the illustrative analysis, acknowledging the sample is much too small for use in real research, where samples are much larger. I consider the question whether occasional shoplifting (Q4)

	Q1	Q2	Q3	Q4	Q5	----------	QJ
R1		4	**5**	0	4		38
R2	1	2	3	**0**	3		56
R3	4	2	3	**1**	4		47
R4	3		**4**	0	3		39
R5	4	4	**3**	1	4		41
R6	2	4	5	**0**	5		34
--							
--							
R*N*	3	4	4	0	4		44

FIGURE 2.3   Fictitious data for the survey containing *J* questions (first five corresponding to Q1—Q5 in Figure 2.2) answered incompletely by *N* respondents. The figure shows only the data for the first five questions and the first six respondents. Score ranges (constructed responses): Q1: 1—5, Q2: 1—5, Q3: 1—6, Q4: 0—1, Q5: 1—6.

relates to income (Q3). Only respondents R1 and R5 have scores (boldfaced) on both Q3 and Q4, so that only their data are usable. Because R5 has the lower income and unlike R1 admits to shoplifting, this suggests that a lower income relates to more shoplifting. Relying on the two complete cases out of six cases in total is hazardous. Let us assume that the four respondents did respond and that by some magic I knew their scores (printed in bold between columns in Figure 2.3, underlined next). R2 provided scores (3, 0), R3 scores (3, 1), R4 scores (4, 0), and R6 scores (5, 0). Then, the three lowest incomes (all scores 3) would involve two shoplifters while the three highest incomes (scores 4, 5) would involve none. This suggests a negative relation between income and shoplifting. If R3 would have had scores (3, 0) and R6 scores (5, 1), income and shoplifting would be unrelated. Of course, I simply cannot know what the missing scores would have been, illustrating that incomplete data are problematic.

Excellent solutions exist for handling missing data, but researchers often use sub-optimal methods that are popular and easy to use (Van Ginkel et al., 2010). An example is leaving out incomplete data lines and then analyzing the remaining complete dataset and publishing the results, with or without reporting how one handled the missing data. The two popular methods are listwise deletion—leaving out all respondent data that contain at least one missing value from all statistical analyses—and available-case analysis—leaving out respondent data only when one needs the data for local computations, as I did in the toy example in Figure 2.3. In listwise deletion, if John did not answer only Q1 about income, the method leaves out his data from all analyses. In available-case analysis, the method leaves out John's data only in all computations that involve income. Both methods produce a smaller data subset resulting in *reduced precision* of the results. In addition, a smaller dataset reduces the probability of finding a phenomenon when it exists (*reduced power*; Appendix 4.1). Worse, leaving out respondents may bias the composition of the data subset, thus suggesting phenomena that do not really exist (*bias*), rendering the research useless.

To illustrate bias, assume that many respondents did not fill out Q1 about income percentage spent on consumer goods. If these were respondents with little education, the sample underrepresents this subgroup. When education level is important for the research question, underrepresentation is damaging to the results, because for the lower-educated respondents, we do not know the percentage of income spent on consumer goods, but we also lost them in relation to the data for the other questions. In general, leaving out respondents with missing scores changes the sample composition relative to the population. Listwise deletion and available-case analysis can be used responsibly only when the sample is large, few cases show missing scores, and missingness is unrelated to the other variables.

Methods superior to listwise deletion and available-case analysis that use all observed data and throw away nothing are available (Allison, 2002; Schafer & Graham, 2002; Van Buuren, 2018). For example, some methods use the association between patterns of missingness for different questions to estimate the missing values from the available data. Other methods model the data and use the model to make inferences about results of interest, such as relations between variables. However, these approaches produce useful results only when the data satisfy specific conditions. Otherwise, the methods produce misleading results. Statistics is like real life: There is no such thing as a free lunch. This is also true for listwise deletion and available-case analysis. If the researcher does not know this and uses a method without having checked the conditions for its proper use, the results are likely flawed. Flawed results can be favorable or unfavorable. Not knowing such things renders research decisions arbitrary and only correct by accident.

When a researcher blindly uses a method trusted by everybody in her research group, and the method produces results that look fine, no one will notice that they are doing something wrong using a method inferior to other methods, under conditions invalid to the method's use, or both. When researchers do not know that they should use the best method available and must check the conditions for adequate method use before using it, they are unaware that they are doing something wrong; hence, there is no intention of wrongdoing. *An interesting question is: When researchers are not statisticians, should they have knowledge about available methods and conditions for the methods' responsible use; at least know that simply applying statistical methods without concern for their adequate use is hazardous; or involve adequate statistical expertise?* It is no exception that some research areas use sub-optimal methods when better methods are available or use certain methods in inappropriate circumstances.

Van Ginkel et al. (2010) studied the incidence of missing-data problems and their handling in personality psychology using questionnaires for data collection. They considered a sample of 832 articles published in three leading journals in the field between 1995 and 2007. They found that about one-third reported missingness on questions. Researchers used twelve methods to handle missingness. Van Ginkel et al. (2010) found that out of 365 attempts at handling missingness, 185 (53%) used listwise deletion, 21 of which first checked whether conditions for responsible application were satisfied. Sixty-five (19%) attempts used available-case analysis, only one of which checked safe-application conditions *a priori*. Remarkably, researchers applied the methods irrespective of the satisfaction of the conditions for safe method use. The authors noted that articles discussed missing-data problems as a nuisance in the data collection rather than a serious problem that threatens the validity of the results, apparently not realizing the impact an incorrect approach to the problem may have. The authors also noted that the popularity and easy application of listwise deletion seems to obscure the view on alternative, statistically superior missing-data handling methods. Application often is an automatism, which discourages using other and superior methods.

DO USING SUB-OPTIMAL METHODS AND INCORRECT METHOD USE QUALIFY AS QRPs?

The outlier example presented a case of p-hacking by tailoring the data after the result was known to obtain a more favorable result. The multiple regression example involved a researcher who trusted a procedure included, for example, in SPSS (IBM Corp., 2021) but did not know that the blind application of the procedure can leave her behind with biased results. The missing data example showed the use of obsolete and inferior methods by a researcher apparently unaware of the availability of superior methods. The outlier example demonstrates a QRP, an observation I expect few readers will challenge. The multiple regression and missing data examples raise the question whether inexpert use of available statistical methods leading to biased or inferior results qualify as QRPs. The issue is whether researchers have a professional obligation to know these things when they use statistics to analyze their data. I discuss this issue in this section.

Interestingly, listings of QRPs (e.g., Bouter, Tijdink, Axelsen, Martinson, & Ter Riet, 2016; John, Loewenstein, & Prelec, 2012; Wicherts et al., 2016) rarely mention the use of sub-optimal (or incorrect) methods or the use of methods when conditions for their use are violated, but they do mention the flawed use of otherwise correct statistical methods. For example, they mention fitting many statistical models and only publishing the best fitting model (notice that of a series of estimated models, one fits best, but what does this mean?). Backward stepwise regression is a fine example of this QRP category. In addition, such QRP lists mostly focus on procedural flaws in data collection, collaboration among researchers, and publication mores. Examples of procedural flaws are repeatedly collecting additional data and in between testing a null hypothesis until significance (data collection, p-hacking), including someone as coauthor who did not contribute (collaboration), and not publishing results that do not match one's expectations (publication mores, file drawer problem).

Bouter et al. (2016) found that participants in a survey presenting 60 instances of FFP and QRPs ranked "delete data before performing data analysis without disclosure" sixth in perceived negative influence on the validity or truth of research results. This category refers to unreported manipulation of data, whereas my outlier example concerned leaving out data after having seen initial results. Although not explicitly mentioned, this category could also include unfortunately deleting cases including missing data by using listwise deletion or available-case analysis. Thus, the category does include using an inadequate statistical method or using a method irresponsibly. John et al. (2012) asked research psychologists whether they were ever involved in one or more out of ten QRPs not including use of inadequate methods or using a method irresponsibly. Wicherts et al. (2016) provided 34 researchers' degrees of freedom concerning arbitrary choices the researcher can opportunistically make to increase the probability of rejecting the null hypothesis or p-hacking, thereby increasing chances of finding results and effects that fail when one repeats the study with new data (also, see Ulrich & Miller, 2020). Examples that come close to inadequate-method use are "choosing between different options of dealing with incomplete or missing data on *ad hoc* grounds" and "choosing between different statistical models." I interpret *opportunistically choosing* as deliberately manipulating the data analysis to obtain a desirable outcome irrespective of the truth. Thus, the question is whether Wicherts et al. (2016) are dealing with QRPs or rather novel instances belonging to the category of FFP due to intention of cheating. Steneck (2006) explicitly mentioned use of improper statistics and data analysis compromising accuracy of research, and Gardenier and Resnik (2002) discussed misuse of statistics in the context of research integrity distinguishing two types of misuse. The first type produces distorted or artificial results (e.g., using listwise deletion when a small sample suffers from excessive missingness and conditions for method use remain unchecked), and the second type fails to disclose important information about statistical methodology to researchers (e.g., not reporting having used listwise deletion).

Although the literature is somewhat ambiguous on the status of use of inadequate methods or using a method irresponsibly, clearly researchers must avoid these practices. There is little if any excuse for using obsolete, sub-optimal, or inadequate statistical methods or using a method irresponsibly, and inquiring minds find out that better options exist even when they lack the skills to apply them. In the latter case, a statistician can offer solutions. Whether a particular instance is a QRP or belongs to the category of FFP (even though it is not included in one of the two Fs and the P) depends on whether the researcher was unaware of incorrect usage or intentionally misused statistics to obtain a desirable result rather than a true or valid result. Individuals deliberately engaging in fraud provide a problem to science but QRPs present a greater problem. In addition, identification of fraud does not reduce QRPs (Fanelli, 2009). We must identify fraudsters, realizing this is easier said than done, and put most of our energy in the reduction of the greater problem of QRPs. Fighting the greater problem does not imply that one should neglect the lesser, albeit serious, problem. From a policy point of view, however, focusing on reduction of QRPs rather than on identifying intentional QRPs qualifying as FFP is more productive for improving the quality of science.

Is it reasonable expecting researchers to know which statistical methods are adequate and how to use them? If a researcher were unaware of the inadequacy of a method, why would she inquire about better methods if she thinks she is doing the right thing? A counter-argument is that a researcher knows she is not a statistician, and that she should seek advice of people more skilled, such as statisticians. This is a lot to ask of people. The researcher may have known that better methods exist but did not expect much of them, found them unduly complex, thought they all come down to the same conclusion, or hesitated to ask a statistician because she refused to appear incompetent. After I gave a lecture at a US university, a PhD student came up to me and told me that her supervisor insisted that his PhD students must resolve their statistical problems without consulting a statistician, because only then would they learn how to solve data-analysis problems and become full-fledged researchers. When I suggested that this could well mean that PhD students made several wrong choices to no avail, she fully agreed but saw no other option than to obey her supervisor's instructions. She still had a career to pursue!

I will drop the issue of intention in case of QRPs and focus on measures that improve the practice of research by preventing QRPs as much as possible and correcting them when they happen. Therefore, the researcher using an inadequate method or who uses adequate methods incorrectly, no matter whether she does this out of ignorance, laziness, embarrassment, or under pressure should receive help to make the right choices. Before I discuss the incorrect use of statistics as a cause of QRPs, in Chapter 3, I follow up on the fraud case of Diederik Stapel as an example of a game-changing event that stimulated a greater awareness of the ubiquitous presence of QRPs in research and the need to do something about them. I discuss policy measures and come back to them in Chapter 7 after I have discussed incorrect statistics use as a cause of QRPs. In Chapter 4, I explain the characteristics of manipulated or manufactured data, and in the Chapters 5–7, I discuss the traps and counterintuitive results of statistics for researchers who are not trained statisticians.

TAKE-HOME MESSAGES

Statistics is difficult. Even if you apply a method correctly, you may find yourself in the end reporting invalid or inferior results.

Trained statisticians still find statistics difficult, so do not worry admitting you find statistics difficult as well. A statistician might also learn from your experiences.

Let no supervisor tell you that the best way to learn is by solving statistical problems without expert help. You learn best from expert examples.

APPENDIX 2.1: METHODOLOGICAL CONCEPTS

I will briefly explain the concepts of research unit, data, variable, sample, and population. This summary serves as a gentle introduction for readers who are not familiar with these concepts but is far from complete. *Research Unit.* In the human sciences, a research unit refers to the entity of interest that provided the data. Other labels for research unit are case, participant, respondent, and subject, the latter three instances always referring to human beings. In addition to being an individual person, a research unit can also be a married couple, a family, a school, a hospital, and a municipality. Married couples can provide information about their consumer habits, families about their social contacts, schools about their learning results, hospitals about their patient service, and municipalities about their public transportation policy. This information is the input for the statistical analysis. A common label for the input is data. In the exact sciences, research units also are very diverse, for example, animals, bacteria, plants taken from different areas, samples of water taken from the sewer system, samples of soil collected from various industrial sites, rocks, elementary particles, and solar systems.

Data. Data can be almost any source of information, such as responses to written questions, answers to oral questions provided to an interviewer, choices from different options as in multiple-choice problems, rankings of objects, documents, demographics, or medical data. The latter category is huge and encompasses answers to questions how someone recovering from a medical treatment gets along at home to enzyme levels and oxygen concentration in one's blood. Data are often qualitative when collected consisting of words and sentences or numerical outcomes of arithmetic exercises but also images from a brain scan. Because such data are difficult to analyze statistically, researchers often transform or *code* the qualitative responses to numbers. The numbers represent meaningful categories in which responses were given—for example, 1 for a correct answer to an arithmetic problem and 0 for incorrect answers—deleting information about the kinds of errors. Other labels for codes are *values* or *scores*.

Sometimes, researchers suspect that quantifying the raw data might lead to the loss of valuable information. One can avoid loss of information by distinguishing different codes for different response types considered relevant to one's research question. Another reason for not quantifying the raw data is the availability of only a few research units. This happens when research units are rare and difficult to find or when collecting data from them is time-consuming. One may think of interviews with CEOs of multinationals or studying bankruptcies of hospitals. Using simple statistics requires data from at least a few dozens of research units and use of more advanced statistical methods requires data from hundreds of research units. Methods are available for analyzing qualitative data (e.g., Ritchie, Lewis, McNaughton Nicholls, & Ormston, 2013).

Variable. A variable is a quantity for which the different research units in a study provide data, and that varies—is variable—with respect to its values. Variables are numerous, and can be questions about consumer behavior, background variables such as age and education level, childcare professionals' ratings of children's activity, playfulness, and social interaction in kindergarten, and intelligence, arithmetic, and language assessments. Researchers collect data for several variables, today often dozens if not hundreds fueled by the availability of wearables and the Internet, and advanced medical equipment such as brain scanners. Statistical analysis always takes place on variables; variation is key to statistical analysis. Without it, statistics is dead.

Sample and Population. Researchers draw samples from much larger populations for two reasons. First, populations are far too large to include in one's study. Second, provided the sample is representative of the population, a relatively small sample can substitute for the population. Studying all 330 million US citizens with respect to some research question not only is too immense a task, but also represents a waste of time and resources compared to studying a sample of, say, a couple of thousand respondents. In terms of information, increasing the sample size is an instance of the law of diminishing returns, and soon adding more respondents provides negligible gain, hence is useless.

The representativeness of the sample for the population is an issue of continuous concern. In many cases, to be informative of a population, a sample must look like it. Let me take research aimed at finding a vaccine against COVID-19 as an example. If the population consists of various age groups, healthy people and people suffering from physical inconveniences, and males and females, the sample must represent all these groups when the researcher has reason to expect the vaccine may not be equally effective or safe among the subgroups. If she would include only younger, healthy men, her results would be valid for only this subgroup, and the sample is not representative of the population. Sometimes subgroups are deliberately oversampled when, for example, the researcher is interested in learning about side effects of a certain dosage of an experimental vaccine and needs high power in the subgroups receiving this dosage.

Statistical theory requires one must draw samples at random from the population. This means that each person in the population must have the same probability of selection in the sample. Suppose the population consists of 5,467,892 persons and one needs a sample of size 20,000; then random sampling is comparable to drawing 20,000 names blindly from 5,467,892 cards out of a big hat. If taken as a hard condition for doing research, the researcher needs a complete list of people in the population and a procedure for random sampling. There are a few practical problems here.

First, if available, lists of population members are usually incomplete. A population is a dynamic entity, members move in and out of it. People are born or they pass away, students start a training and others drop out or finish it, and healthy people contract a disease while patients recover. Second, populations are sometimes difficult to define. For example, when one performs a study in the US population, does this include only people with the US nationality, all people legally living in the US, all people living in the US including those without a residence permit, any of these groups living there in 2020, and so forth. Assuming one has adopted a definition and has access to a reasonably complete list, then drawing research units at random means losing control over the composition of the sample, hence its representativeness of the population. The effect is smaller as the sample is larger, but smaller subgroups may be absent or nearly so. If the goal of the study is to have a representative overall sample, one might isolate all the relevant subgroups first and then sample at random from each separate subgroup a subsample of a size proportional to the subgroup size in the population. If the subgroup itself is important in the study, one might do a separate study in the subgroup.

Oddly enough, sampling from meaningful populations is an underrated topic in much research with human beings. Often the researcher uses convenience samples. These samples happen to be available but may not be the result of random sampling, let alone that they represent the population well. Thus, statistical analysis may be questionable, and generalizing results to a population is problematic. Without the possibility of generalization, what is the point of doing research anyway? This is a greater problem in much of the human sciences than many people realize, and I advise the reader to check the "Method" sections in many articles across a wide range of journals. My experience is that researchers often describe their samples rather well in terms of background variables, such as gender, age, education, and so on, but that a description of the population to which one wishes to generalize research results is poor if not absent altogether. Mellenbergh (2019) discusses several sampling strategies common to the human sciences.

APPENDIX 2.2: MULTIPLE REGRESSION MODEL

I denote predictor j by notation X_j, where X is the variable representing the score and the subscript j enumerates the predictors $j = 1, ..., J$. The criterion is Y. The multiple regression model weights the predictors by their importance to predict the criterion. The regression weights are denoted by the Greek lower-case beta as $\beta_1, ..., \beta_J$. It is common that a set of predictors does not perfectly predict a criterion. The discrepancy or the residual between observed and model-predicted criterion scores is denoted by the Greek lower-case epsilon, ε. The multiple regression model for respondent i equals

$$Y_i = \beta_0 + \beta_1 X_{i1} + ... + \beta_J X_{iJ} + \varepsilon_i.$$

Parameter β_0 is the intercept, which has little relevance for the example discussed in this chapter and is further ignored. The multiple regression model is estimated from the data, an exercise that is routine in statistics but beyond the scope of this book. The sample model looks like

$$\widehat{Y}_i = b_0 + b_1 X_{i1} + ... + b_J X_{iJ},$$

where lower-case b stands for the weight estimated from the sample data and $\widehat{Y}_i$ is the estimated criterion score. The difference $\widehat{\varepsilon}_i = Y_i - \widehat{Y}_i$ is the estimated residual.

One can use different sets of predictors to predict criterion, Y. The degree to which a set of predictors is successful predicting a criterion is expressed by the squared correlation between the observed criterion scores in the sample, Y, and the criterion scores predicted by the multiple regression model using the observed predictor scores, $\widehat{Y}$, and is denoted $R^2 = r_{Y\widehat{Y}}^2$, where r is the sample correlation. The correlation, $r_{Y\widehat{Y}}$, expresses the strength of the linear relation between the observed and the model-based criterion, and the squared correlation expresses the proportion of variance the estimated model explains of the variance of Y, denoted S_Y^2 (squared upper case S^2 for sample variance). This correlation is the multiple correlation. The variance of residual $\hat{\varepsilon}$ across respondents in the sample, $S_{\hat{\varepsilon}}^2$, is a measure of the failure of the model to predict criterion Y: The greater the residual variance, the larger the part of Y the model does not predict. One can use statistical testing to establish whether sample weight b provides evidence that the corresponding weight β is different from zero.

NOTES

1 https://www.merriam-webster.com/thesaurus/erratic
2 https://www.collinsdictionary.com/dictionary/english/erratic
3 https://www.merriam-webster.com/dictionary/accident
4 https://www.collinsdictionary.com/dictionary/english/accident
5 Welch correction means assuming different variances in income in the two neighborhoods. This is reasonable because we do not know in advance whether variances are equal or unequal.
6 I loosely define the human sciences to include anthropology, economics, education, health research, policy research, political science, psychology, sociology, and other disciplines that focus on human behavior. Of course, FFP and QRPs are also common in the sciences and the humanities, as a rich literature in these areas sadly witnesses.

CHAPTER 3

Learning from Data Fraud

I N THIS CHAPTER, I follow up on the Stapel fraud affair as a con-
crete case of FFP providing a point of departure for thinking
about policy measures aimed at prevention of FFP and QRPs discussed in this chapter and Chapter 7, and examining features of
unrealistic data, discussed at some length in Chapter 4. I also
spend some time on the question why fraud cases tend to remain
undetected for such a long time. This is perhaps the most specula-
tive part of the book, but unavoidably so and a topic of great
urgency, thus deserving ample attention to increase further
awareness of the imperfections in research.

HINDSIGHT BIAS

In the next 15 months until the final report came out, the Stapel
affair was prominent in the media and hardly allowed several others and me, who had to deal with its consequences, a moment of
rest. At first the media directed attention at Tilburg University, the
initial crime scene. Unavoidable questions were whether we had

had any suspicion and whether we could have known what Stapel was doing, had we paid more attention. The equally unavoidable answers were that we had not had a clue and now felt as trapped and cheated as anyone else felt. I noticed this is a difficult topic for many people, mainly because they find it hard to eliminate the knowledge available after the incident and ask whether in all reason one could have known what was going on before the incident when that information was unavailable. Looking back, surely there were signs, but these signs derived their meaning from the context in which they happened. It makes all the difference in the world whether that context is the situation before or after the truth came out.

The question of whether we should be more critical of our colleagues' work and them of our work and if so, how we can implement such a behavior change undoubtedly affecting our professional relationships, is an interesting one. For a couple of misconduct cases in other universities that became public after the revelation of Stapel's fraudulent activities, the duration of the misconduct lasted a long time and ten to 15 years is no exception. Here are four such misconduct cases.[1]

- Eyewitnesses and documents the Vrije Universiteit Amsterdam anthropologist M. M. G. (Mart) Bax[2] claimed to have used in two of his research projects taking place between 1974 and 2002 could not be retrieved, justifying the conclusion that reasonable doubt about their existence was warranted (Baud, Legêne, & Pels, 2013). In addition, the committee overseeing the Bax case found that Bax republished several of his publications with minor changes but without telling while other publications could not be retrieved thus appearing non-existent, showing a way of working generally lacking transparency and verifiability. The committee noticed a late twentieth century organizational culture of small and isolated research groups whose work quality evaded external assessment, whereas today they are driven more by collaboration and acceptance of mutual criticism (Baud et al., 2013, pp. 42–47).

- In approximately the same period as Stapel was active, Erasmus University Rotterdam's Don Poldermans,[3] specializing in internal and cardiovascular medicine, failed to obtain informed consent from patients included in research and reported permission from medical ethics committees when such permission, even though legally unnecessary, was absent. In addition, data collection, data manipulation, and results contained inaccuracies, coauthorships were assigned inconsistent with organizational regulations, and data were reported more complete than reality justified, thus opening the possibility of biased results. On page 6 of their advice, the Integrity Committee noticed a research culture of relative indifference, and on page 8, they recommended fostering shared responsibility of research integrity.[4] Poldermans' case is different from several others discussed here in that his tainted results may have had adverse impact on the success of medical interventions (Bouri, Shun-Shin, Cole, Mayet, & Francis, 2014; Cole & Francis, 2014a, b; Lüscher, Gersh, Landmesser, & Ruschitzka, 2014). The other cases discussed here, including Stapel's, may be seen as contributing negatively to theory building or policy development but without direct consequences for individuals (except, of course, colleagues involved without knowing).

- Psychologist Jens Förster,[5] affiliated with the University of Amsterdam, reported linear trends in three articles published in 2009, 2011, and 2012 that were highly unlikely in real data without unwarranted manipulation. In addition, he reported data not containing any missing data, experiments without participant dropout, and participants not expressing awareness of the deceit usually exercised in experiments, events unlikely in psychological experiments leading to suspicion of data authenticity. The Department of Psychology took several measures to improve policy with respect to research, noticing that researchers still were not aware enough of the need of transparent data storage.

• More recently, between 2012 and 2018, Leiden University psychologist Lorenza Colzato[6] did not include deserving coauthors on publications, failed to obtain permission from a medical ethics committee for taking blood samples (a criminal offense under Dutch law), manipulated data to obtain favorable results, and reported research in grant applications that likely never took place. On page 26 of their report,[7] the Integrity Committee that reached these conclusions implied that "the appropriate channels for early detection of possible violations of academic integrity did not function." A second report was published in 2021,[8] identifying 15 tainted articles in total. Often, subjects providing data inconsistent with the researcher's expectations were deleted from the analyses. Original data for 27 articles were no longer available, rendering further investigation impossible.

Scientific integrity committees looking into alleged fraud cases and advising universities' executive boards usually mention the organization culture of insufficient shared responsibility as the culprit of the problem and offer suggestions for improvement. The committees investigating the long-lasting Stapel and Bax cases also noticed that individuals suspicious of a colleague's misconduct experienced restraint taking steps. Organizational obstacles will surely play an important role, but taking individual responsibility or refraining from it is another issue. The Stapel case showed that three junior researchers prepared their case as well as they could and then took a bold step forward. However, that was only after 15 years in which no one else undertook decisive action. Why did the revelation of these and other cases that are never one-day incidents take so long? Did no one have any suspicions? I think they did, but that in general it takes a giant step for an individual to act and report one's suspicions to a counselor in charge of research integrity or a supervisor. Here are some reasons.

First, one may prefer to be certain about the allegation and wish to avoid a false claim and unjustly damage a colleague's reputation. While in some cases such *prudence* proved unwarranted afterward, calls for low-threshold complaints procedures sometimes seem to ignore that the defendant may be innocent and that an unfounded complaint may cause a lot of harm. People who were falsely accused suffer sleepless nights just like people who found themselves the victims of fraudsters and may take long to overcome the negative experience. Second, one may find it hard to believe that a colleague thought to be respectful would engage in serious misconduct and thus considers what could be a misstep as an unintentional error one should preferably ignore. I think people tend to *deny* such inconsistencies or at least rationalize one's responsibility for taking steps, preferring life to continue with a minimum of disturbances. Third, hierarchical inequalities may play a role when the potential complainant decides the risks are too big. In such cases, people are simply afraid they do not stand a chance when filing a complaint due to mistrusting authorities. They *fear* a complaint might turn against them. Fourth, in the midst of the Stapel case, one of the guests at a dinner party confided in me that when he would find evidence of fraud in a manuscript he reviewed for a journal, he would keep the knowledge to himself. He simply did not want the hassle. People are *reluctant* or even *unprepared* to take the risk of filing a complaint and do not want to get involved in a situation that takes a lot of time and energy and perceive is not their business anyway.

Prudence, denial, fear, and reluctance may be some of the likely motives for passivity, and the bystander effect known from social psychology provides a fifth motive, referring to the phenomenon that an individual tends to be passive toward a victim when other people are present. The effect is that misconduct that several people have heard of, suspect, or have seen with their own eyes and perhaps even talk about at the water cooler drags on for too long.

In the context of research fraud suspicion, the individual may easily decide that it is the supervisor's, manager's, or administrator's responsibility to act. When many individuals decide to stay passive bystanders, the consequence may be that the supervisor, manager, or administrator remains ignorant and thus cannot act. The three whistleblowers in the Stapel case found his results were too good to be true, and in the previous 15 years more colleagues must have had similar thoughts without pushing them to a next level. The three committees investigating Stapel's misconduct concluded that several senior colleagues remained passive before the three young whistleblowers acted.

Nowadays, Dutch universities have all embraced the Netherlands Code of Conduct for Research Integrity (KNAW; NFU; NWO; TO2-federatie; Vereniging Hogescholen; UNL [previously VSNU]),[9] 2018). They all have assigned counselors for research integrity that researchers and other members of the academic community can consult confidentially in case one suspects a colleague of a breach of integrity, and have installed committees that investigate such alleged breaches of research integrity when someone files a complaint. The legal rules under which these committees do their work ensure that the complainant and other interested parties are protected and do not suffer undue disadvantage in their career prospects or otherwise. I have noticed through the years that, even under these terms, potential complainants still do not trust the system enough to file a complaint or fear the defendant might come after them. An alternative strategy could be not to file complaints alone but with a small group or include a supervisor who can take the lead. The three whistleblowers in the Stapel case followed this strategy, but often one is alone facing a difficult decision. Unavoidably, it takes some degree of *civil courage* to raise one's hand, trusting that one is doing the right thing, but the desire to have complete certainty about the allegation, disbelief that someone engaged in wrongdoing, fear of the consequences of filing a complaint, and reluctance to get involved may drive one to doing nothing. The fact that Stapel was an influential scientist, highly regarded by many colleagues, and believed to enjoy protection by people in powerful positions perhaps slowed down potential whistleblowers, a conclusion the committee looking into the Stapel case inferred as well.

THE STAPEL CASE AS A CATALYST REDUCING MISCONDUCT

Terminating Stapel's contract and installing the Levelt Committee, asking them to turn every stone to find the truth, was all the university could do in the first few days. It also became clear quickly that Stapel's breach of integrity started much earlier in Amsterdam and came to full bloom in Groningen before it reached Tilburg, where it blossomed further and was finally unmasked. Because of these reasons, within a few weeks the discussion in the media took a different turn and moved to the question of whether Stapel was a unique case and how universities in general dealt with suspicion or accusation of alleged breaches of misconduct. Although Stapel's name turned up in every newspaper article and news flash concerning research integrity, the media now asked many questions about the way in which scientists conduct research. Also, many new cases of scientific misconduct in other universities were reported, bringing to light that Stapel's was not unique, albeit extremely serious, given its scale and the number of coauthors and PhD students whom he implicated in his practices simply by working together with them or supervising them.

For each journal article, the Drenth (University of Amsterdam), Noort (University of Groningen) and Levelt (Tilburg University) Committees investigated whether it was tainted using manipulated data, made-up data, or fantasy—research that never took place supposedly using data that never existed. The three committees concluded that coauthors and PhD students simply trusted their colleague or mentor to be of goodwill and high integrity and were therefore misled. Again, looking back, readers may be inclined to think that there must have been people noticing a swindle so big before the whistleblowers revealed their findings.

Again, I remind those readers that when you do not have a clue that someone is cheating you, you do not have reason to think you are being misled and will behave accordingly, acting as if everything is normal. The cheater who knows exactly what he is doing has all the time and opportunity in the world to lead you astray and adjust his actions when deemed necessary to prevent you from becoming suspicious. Then, working together with a successful colleague or receiving supervision from a famous professor is attractive, and why should it not be?

The attention the media gave to scientific research and research integrity caused a quickly growing awareness in researchers and administrators that science could not return to business as usual. The Stapel affair was a catalyst in this process, not the cause (Huistra & Paul, 2021). Before Stapel, other integrity scandals saw the light, and this has been the case for as long as science exists, in and outside the Netherlands. Broad and Wade (1982), Judson (2004), Goodstein (2010), and Stroebe, Postmes, and Spears (2012) provide overviews of older cases. Kevles (1998) discusses the Baltimore case from the field of genetics, and Craig, Pelosi, and Tourish (2020) discuss the case of psychologist Hans Eysenck, both cases rather controversial. Van Kolfschoten (1993, 2012) discussed overviews of dozens of cases at Dutch universities. For example, in 1990, Henk Buck (Maddox, 1990; Van Kolfschoten, 2012, pp. 169–180), a professor at Eindhoven University of Technology published an article in *Science* (Buck, Koole, Van Genderen, Smit, Geelen et al., 1990) claiming he had found a cure for AIDS, ignoring concerns expressed by several colleagues before and after publication that the lab results were probably contaminated. Two committees looking into Buck's work inferred an intimidating work climate at Buck's lab and indications of fraudulent behavior, inferences the university first accepted and withdrew later. The *Science* article was retracted. Buck retired a few years before the official date. Another example concerns René Diekstra (Bos, 2020, pp. 75–78; Van Kolfschoten, 2012, pp. 62–80), a professor of psychology at Leiden University, whom in 1996 some media and colleagues accused of plagiarism in popular and scientific publications. After a committee investigated the case and confirmed the accusations, Diekstra resigned. Both the Buck and Diekstra cases got intensive media coverage and suffered from a long aftermath of bitter controversy, but neither had the impact Stapel had.

The first time I heard of fraudulent behavior in science was in 1989, when my former PhD supervisor, Ivo Molenaar, a professor of statistics at the University of Groningen, told me confidentially about an applicant they interviewed recently and offered a job, and then found out might be a fraud.[10] In fact, what had happened was that Tom Snijders, another Groningen statistician and member of the interview committee, accidentally ran into an article on a statistical topic that was published in a psychology journal and that looked very much like his master's thesis that he completed a couple of years ago and later published in a statistics journal. Several text fragments, mathematical equations, and a table were literally equal to their counterparts in Snijders' master's thesis and article. One of the authors of the alleged plagiarized work was the applicant, Johannes Kingma. The article based on Snijders' work did not appear in Kingma's vita he sent to the interview committee. Molenaar's reading of the publications convinced him that the similarities were more than accidental, and he informed the interview committee's chair of the apparent fraud. The committee asked Kingma for an explanation. Kingma declared he did not know Snijders' work and that his article was original work. The committee was dissatisfied with this explanation and withdrew their job offer. Kingma did not accept this and went to court, demanding a new investigation on the alleged plagiarism. This unusual step had the effect of the court asking two expert statisticians for their assessment. They reached the same conclusion Snijders and Molenaar reached earlier, and Kingma's hubris became his downfall. During this episode, I also learned about the effects this kind of impertinent behavior—not only the plagiarism but also the nerve to go to court—has on an organization and the people working there. My supervisor was upset about the incident and suffered several sleepless nights before the University was able to terminate the Kingma case.

Although Stapel did not introduce fraud in science—far from it—the question is justified why the Stapel case attracted so much attention from the media and colleagues from all over the world (e.g., Budd, 2013; Callaway, 2011; Chambers, 2017; Chevassus-au-Louis, 2019; Craig, Cox, Tourish, & Thorpe, 2020; Haven & Van Woudenberg, 2021; Markowitz & Hancock, 2014; Nelson, Simmons, & Simonsohn, 2018; Stricker & Günther, 2019; Zwart, 2017). Sure, it was outrageous in its scale and depth. The fraud lasted for at least 15 years and misled dozens of colleagues, and most seriously, several PhD students who saw their theses go up in smoke when the fraud came out. The unscrupulousness of involving PhD students unknowingly and undermining their work by providing them with falsified or fabricated data when they thought the data were real and stemming from experiments with real people was perhaps the most shocking to colleagues and the public. The Drenth, Noort and Levelt Committees came up with another finding they called sloppy science, like the QRPs I discussed in Chapter 2, that may have rung bells that many researchers preferred not to hear. The committee's findings were not new but had an impact on Dutch universities that no one could ignore. In the next years, universities sharpened their integrity policy and had employees sign for consent. They provided courses on research integrity and good research practices, expanded the reach of research ethics committees assessing research proposals, and encouraged or even forced their researchers to save data packages including raw data and processed data (e.g., resulting from missing data procedures) on institutional or nationally available servers. They also sharpened promotion regulations describing conditions for PhD research and requirements for PhD theses.

Universities implemented all of this and even implemented more measures, but some researchers and research areas still are in the process of realizing complete acceptance. A reservation heard in response to the measures is that researchers feel and object that administrators do not take them seriously or even mistrust them. Other researchers think that the way they do research is their responsibility and not that of the Dean or the University's Executive Board. To amplify this argument, some researchers appeal to *academic freedom* as the ultimate safe haven for the scientist, protecting her from the interference of non-scientists. Academic freedom has received much attention in the literature, and I do not claim I have the ultimate wisdom with respect to this admittedly difficult topic. Therefore, I will only say a few words about it based on my experience in academia. They amount to the opinion that academic freedom is a highly important attainment but does not provide absolute freedom to the academic to do whatever she pleases. Academic freedom means that scientists are free to investigate whichever topic they choose and can do that without the interference of the State, politicians, religious movements, the University, and fellow scientists, and any other institution or group I forgot to mention. Put simply, in science we do not recognize any taboos on what we investigate. This is a principle that, like all principles, is under pressure as soon as other interest groups start showing curiosity.

A now notorious and admittedly extreme example of the containment of academic freedom is Dutch criminologist Wouter Buikhuisen's 1970s research on biological correlates and determinants of criminal behavior. In that era, the belief in nurture as opposed to nature rendered the idea that behavior is partly biologically determined unwanted. This negative attitude was based on a particular view of the world and humanity but lacked a scientific basis. As a result, not only his research but also Buikhuisen himself became the target of fierce and aggressive criticism by newspapers and TV stations, and fellow scientists as well. This controversy finally resulted in his departure from the university and from science in the late 1980s. In the 1990s, brain research became popular in psychology and other science areas, and the taboo on physical determinants on behavior seemed to have never existed. In the new millennium, Buikhuisen was rehabilitated,[11] but too old to pick up his research where he left it behind, if he ever had the ambition.

One does not do her research in total isolation from one's environment. This means that colleagues will have an opinion about the relevance of one's research and are entitled to have such an opinion; this is also academic freedom. The point is that a discussion about the quality of research must use objective arguments and logic, but not reflect political viewpoints, personal beliefs, and likes and dislikes concerning the researcher whose work is under scrutiny. An open discussion in science is as the air that we breathe and need to live, and with discussion comes the readiness of all discussants to accept reasonable arguments that favor or disfavor their position. The utmost consequence can be that the researcher cancels a planned project or at least seriously changes it to accommodate the comments colleagues provided.

Another reason not to realize an initially preferred research plan is that no one finds it of interest. Perceived lack of contribution to solving a particular greater problem or the development of a research area may play a role. Another possibility is that many colleagues tell the researcher that she plans to do something that others already did before. Trying to prove the Earth is a sphere is somewhat of a lame example. A better example is that it happens often that a researcher picks up a topic that another researcher addressed several decades ago and that vanished with the generations that knew the result, not realizing she is repeating the research in a somewhat different manner without adding anything essential. A more prosaic reason is that the researcher's School does not want to finance a particular research plan, because it does not fit into the research programs the School and its academics agreed upon for financing. Interested readers interested in the topic of collegial criticism as *organized skepticism* may want to consult Merton's seminal work (Merton, 1973).

ALTERNATIVE MEASURES FOR PROMOTING RESPONSIBLE CONDUCT OF RESEARCH

A measure that was somewhat unconventional when the Tilburg School of Social and Behavioral Sciences introduced it in 2012 in response to the Levelt Committee's preliminary findings was to install the Science Committee.[12] What is it, what does it do, and why? The Science Committee is an audit committee that assesses and advises researchers' data handling. The name of Science Committee was a euphemism chosen to circumvent possible resistance by academics when held accountable for their research and their data management that existed among researchers of the School shortly after Stapel's unmasking. The reader may find that upon first reading, the previous sentence is a little odd, requiring explanation. Of course, every researcher is responsible for her research with respect to the quality of the theorizing, the research design, the statistical analysis, and the reporting of the results in articles in the professional journals and other media. What the Stapel affair revealed was that, when determined to do so, a researcher could falsify his data without colleagues noticing for a long time and when caught, in one stroke undermine science's credibility. The system's foundation was trust that this would not happen, that science was above this kind of misconduct, and that scientists were infallible messengers of the truth. The long history of science had shown repeatedly that fraud happens—think of heredity researcher Georg Mendel offering an early instance of data too good to be true in the 1860s (Judson, 2004, p. 55), and intelligence researcher Cyril Burt publishing three-decimal correlations that remained the same across non-peer-reviewed publications in the 1950s and 1960s (Judson, 2004, p. 94). However, belief in science's moral purity always seems to prevail until the next scandal that was too big to ignore, and Stapel's fraud seems to present such a landmark.

Still, it took time to change the perspective from complete trust to accountability. I remember a national meeting in Tilburg a couple of months after Stapel's fall where several speakers reminded the audience that "we are not like Stapel," and although I agreed with the literal content of that message, I also noticed a general reluctance to admitting that we could no longer be naïve when it came to accountability. In a critical national newspaper article,[13] a psychologist reminded the nation that some psychological researchers kept the data from their research in an archive box at home, a practice he claimed we could no longer tolerate. Agreeing with him, when I cited his viewpoint in passing during a formal meeting using the archive box metaphor, little did I suspect that the university's newspaper would quote me. This provoked several reactions by researchers who were upset simply by my citation of someone else's article. To calm the emotions, I felt obliged to write a letter to the department heads explaining to them what I meant precisely and, fortunately, this cooled them down.

The incident showed how itchy colleagues had become in response to the Stapel affair. Disbelief, anger, and shame took possession of many colleagues in and outside the School, and also in and outside the University. The fact that a recently so-much-respected colleague, who was even given confidence to be the School's Dean, had fooled and betrayed everybody, was difficult to grasp. Several people had become uncertain about their way of working and even their careers. Others, especially when their organizational unit was remote from the Department of Social Psychology, the School of Social and Behavioral Sciences, or Tilburg University, not seldom expressed attitudes ranging from "Not my problem" to "This would never happen with us." I witnessed and as Dean underwent such painful events in public as well. For example, at one official occasion at another university, in front of a full auditorium, one of the speakers suddenly started fulminating against Tilburg University, apparently unaware of the fact that the Stapel affair was a problem of three universities, Tilburg being the university that finally caught him. Later, appreciation for the way Tilburg University handled the situation replaced initial utterances of frustration and aggression, but it took a while for this to happen. This was the atmosphere in which, only half a year after the Stapel affair took everybody by surprise, we installed the Science Committee. Perhaps a more prudent Management Team would have waited a little longer, but we felt that we should act quickly and do something, taking advantage of the urgency felt by many people and not wanting to waste the opportunity a good crisis offered.

The Science Committee's procedure is the following. Each year, the Committee samples 20 empirical-research publications out of the set of articles researchers from the School publish in national and international journals, chapters in books, and books and PhD theses. For each publication, the Committee assesses the *quality of the data storage* including the availability of a data package plus meta data (what was each author's contribution to the research?) and handling of privacy and retention period of the stored data, and the *reporting of research methods*. Based on the information they collect, they interview the article's first author and provide their opinion about the quality of the data storage and reporting, later adopting the FAIR principles for Findability, Accessibility, Interoperability, and Reusability (Wilkinson, Dumontier, Aaldersberg, Appleton, Axton et al., 2016) that had become available in the meantime. If the Committee comes upon details or greater problems, they *advise* the first author and her colleagues about improving their data storage, the completeness of the stored datasets, honoring subjects' privacy, access to the database in which they stored their data and auxiliary information, and data availability to other researchers. The aims are the following. First, the Committee wishes to encourage a concerted effort to improve *accountability* for data handling and methods reporting. Second, they wish to create an opportunity for all to *learn*. All those colleagues fearing a witch-hunt for data management inconsistencies could rest assured. Third, the Science Committee aims at contributing to the development of the university's data policy and perhaps a national protocol concerning data archiving by researchers in social and behavioral sciences and does this by providing an example.

The installation of the Science Committee was an expression of the sudden awareness that scientists have a responsibility to their colleagues, their universities, to society, and to science. With this responsibility comes the preparedness to be accountable for what one does and how one does that. It is remarkable that it took so long before something as evident as transparency was recognized and put into action by the Science Committee. This is the more remarkable because Stapel's was not the first integrity breach that happened in the long history of science, and the Stapel affair rather was a catalyst that ignited the way scientists thought about subjecting their data management to scrutiny by their colleagues. The resistance several researchers showed initially is understandable in a culture that allowed them to follow their own data policy entirely. In addition, in the context of a huge data fraud case, others might easily take the introduction of an audit committee as a sign of mistrust toward the scientific community. It is easy to say it is not, but will people believe this? I would say that it is in everybody's best interest that, given the persistence of integrity breaches and QRPs in various research areas, obligatory audits, no matter how they are implemented, provide the scientific community with evidence they are aware that they are no saints and take precautions to prevent mishaps as much as possible. In addition, it is important to realize that, in the absence of some obligatory audit mechanism, people like Stapel were able to work in relative isolation, keeping everyone potentially threatening at arm's length. An obligatory audit might scare away potential fraudsters or at least discourage them to a point that FFP becomes a non-profitable "research" strategy.

Does the Science Committee function well? The latest experiences are that some research groups have their data policy better in place than others do; when called to present the data and method information for an article, researchers tend to arrange their data storage only in prospect of an audit; and when people have left the School, they tend to lose commitment. Although there is still much work to do and several researchers still must get used to keeping track of their data management rather than to see it as a nuisance, having the Committee around creates greater awareness, a stronger sense of responsibility, and an increasing preparedness to be held accountable.

Finally, the Science Committee's procedure certainly was not the only way to set up an audit and could be amended easily. The original procedure was less detailed than it is today and was streamlined through the years as practical experience accumulated. As an alternative not implemented at Tilburg University, by the way, rather than a random selection of researchers, which might be interpreted as a deterrence more than a learning experience, one could choose to select for each researcher one publication at fixed and known time intervals and use this for learning opportunities tailored to individuals' personal needs. Of course, such a procedure would be time-consuming, which would need to be weighed up with the expected gain in learning. Labib, Tijdink, Sijtsma, Bouter, Evans et al. (2023) discuss the case of the Science Committee from a governance perspective as an attempt to combine network processes with bureaucratic rules to establish an accepted approach to foster research integrity.

TAKE-HOME MESSAGES

FFP cases can linger on even when colleagues have noticed something odd. Many psychological mechanisms prevent colleagues from raising their hand, but a joint effort can help to remove such barriers.

Because trust is essential for a good research climate, policy measures aimed at reducing risk of derailment may be preferred to individual detective work. Policy measures explicate the rules and affect everyone, not being at the expense of mutual trust in the workplace.

Most scientists value their academic freedom, and they should, but academic freedom does not relieve them from accountability of their research to their colleagues, employer, society, and science.

NOTES

1 The names of the researchers involved have been published abundantly in the media, readily available from an internet search and are therefore not withheld here.

2 Bax' identity is available from the official report by Baud, Legêne, and Pels (2013). See also Vrije Universiteit Amsterdam weekly *Ad Valvas* for a full exposure of his identity, for non-Dutch readers unfortunately in Dutch: https://www.advalvas.vu.nl/nieuws/ 'ernstig-wetenschappelijk-wangedrag'-van-antropoloog-bax

3 Identity explicit from several articles published in Dutch and international journals (Bouri, Shun-Shin, Cole, Mayet, & Francis, 2014; Cole & Francis, 2014a, b; Lüscher, Gersh, Landmesser, & Ruschitzka, 2014; Smit, 2012).

4 Advice Committee Research Integrity Erasmus University Rotterdam to the Executive Board of the University, dated 16 November 2011, short version, Dutch language. Retrieved from the Internet, but non-retrievable on 23 November 2021.

5 Identity explicit from articles in *Science*, retrieved from the Internet on 7 July 2022 (https://www.science.org/content/article/no-tenure-german-social-psychologist-accused-data-manipulation) and *Parool*, a well-known Amsterdam area newspaper, for non-Dutch readers unfortunately in Dutch, retrieved from the Internet on 7 July 2022 (https://www.parool.nl/nieuws/weer-fouten-gevonden-in-publicatie-fraudeprofessor-uva~b3b4622b/?referrer=https%3A%2F%2Fwww. google.com%2F)

6 See Leiden University weekly *Mare* for a full exposure of her identity: https://www.mareonline.nl/en/news/psychologist-committed-fraud-in-15-articles-how-test-subjects-kept-disappearing/

7 Advice Academic Integrity Committee Universiteit Leiden, Case CWI 2019-01, to the Executive Board of the University, dated 11 November 2019 (English translation, 27 pages). Retrieved from the Internet, but non-retrievable on 23 November 2021. Dutch-language version downloaded on 23 November 2021 from: https://www. universiteitleiden.nl/advies-cwi-2019-%2001-geanonimiseerd.pdf, inference with respect to research culture on page 28.

8 Retrieved from the Internet on 7 July 2022: https://www.organisatiegids. universiteitleiden.nl/binaries/content/assets/ul2staff/organisatiegids/ universitaire-commissies/cwi/cwi-20-02-advies-def_redacted.pdf

9 KNAW: Royal Netherlands Academy for Arts and Sciences; NFU: Netherlands Federation of University Medical Centres; NWO: Dutch Research Council; TO2-federatie: Collaboration of Applied Research

Organizations; Vereniging Hogescholen: Netherlands Association of Universities of Applied Sciences; UNL (previously VSNU): Universities of The Netherlands. Note: These are the best translations I could find. If not entirely correct, they characterize the organizations correctly.

10 After having written this paragraph from memory, I checked it with Van Kolfschoten (1993, pp. 50–52) and asked Tom Snijders, and this led to a few minor changes.

11 According to the Dutch national newspaper *Algemeen Dagblad* in 2009; see https://www.ad.nl/wetenschap/affaire-buikhuisen-voorbij-wat-leiden-betreft~af0747a9/

12 https://www.tilburguniversity.edu/research/social-and-behavioral-sciences/science-committee

13 *NRC Weekend, Wetenschap* (Science), page 9, Saturday 30 June and Sunday 1 July 2012.

CHAPTER 4

Investigating Data Fabrication and Falsification

THE LEVELT COMMITTEE WAS installed on Friday, 9 September 2011. This was only two days after Tilburg University informed the public of Stapel's data fraud. Translated from Dutch, the Committee's tasks were:

1. The committee will examine which publications are based on fictitious data or fictitious scientific studies and during which period the misconduct took place.

2. The Committee should form a view on the methods and the research culture that facilitated this breach and make recommendations on how to prevent any recurrence of this.

On 31 October 2011, less than two months later, the Levelt Committee published their interim report, titled (translated from Dutch) *Interim Report regarding the Breach of Scientific Integrity Committed by Prof. D.A. Stapel* (Levelt, 2011). The interim report

DOI: 10.1201/9781003256847-4

discussed the preliminary results of the Committee's investigation based on interviews held with a few dozen colleagues whom the Committee had invited or who stepped forward to provide information, and who had been involved with Stapel in the previous years as a close colleague, a coauthor, a PhD student, or as an administrator. The Committee had collected all Stapel's published journal articles covering his Tilburg University period, 2007–2011. Importantly, the Committee had also invited people who had worked together with Stapel to make datasets available that they or Stapel had used for analysis and reporting, and all questionnaires used, hypotheses formulated, and email exchange between Stapel and them. Stapel provided a list of his publications from the period 1994—2011, ranging from his PhD work at the University of Amsterdam to his professorships at the University of Groningen and Tilburg University, that were based on fictitious data. Soon, he stopped being cooperative. The interim report also contained preliminary conclusions from the Drenth Committee (University of Amsterdam) and the Noort Committee (University of Groningen). I have based part of this chapter on the Dutch-language version of the interim report (I am not aware of an English-language version), and those readers mastering the Dutch language can find the complete report on the Internet.[1] Here, I will focus on the Committee's preliminary conclusions and recommendations.

SOME OF THE COMMITTEE'S FINDINGS

Given two tasks, the Committee's work served three purposes. First, the Committee aimed at giving an indication of the extent of the fraud. They concluded that several dozens of articles used fictitious data or otherwise manipulated results, and that the fraud started in 2004 or earlier. The Committee concluded that further research was needed to identify which articles were tainted and which were not. A complete list would allow journals retracting these tainted articles and providing them with a red flag signaling readers they could not trust the content of the article and should ignore it. In addition, coauthors could be cleared of possible claims of culpable ignorance or even complicity. The Committee concluded that they had found no evidence of these two reproaches—culpable ignorance or complicity—and that Stapel's fraud was a one-man enterprise, successfully misleading coauthors and PhD students.

Second, the Committee studied the nature of the fraud. They established that Stapel manipulated real data in two ways. The first way was that he changed numerical scores in real data, called falsification, with the aim of manipulating results from statistical analysis to be better in accordance with the results the author wished to obtain. The second way was that he augmented sets of real data with data records of real individuals that he copied so that they appeared multiple times in the dataset, suggesting they represented data from different individuals. Is this falsification or fabrication? I think the latter, because one does not change observed scores into other scores that are convenient for one's purposes but creates new data records. Because one person can contribute only one data record, I consider copies fabrications. Stapel also added fabricated data suggesting they originated from real subjects when in fact the subjects were non-existent and added those fabricated data to real-datasets. The effect of these activities was a larger sample size. Larger samples contain more information and make it easier to reject a null hypothesis when it is untrue at the advantage of an alternative and preferred hypothesis, even when the true effect is small and not of interest (Appendix 4.1). Stapel frequently fabricated complete datasets; although reported as stemming from a sample of real respondents, in fact no living respondent ever produced one score and samples of real persons never existed. All data were manufactured with Stapel sitting behind a desk and were a figment of Stapel's imagination. We will see later that manufactured data, created by either changing individual scores, copying data records, or making up data, when analyzed often yield statistical results that are extremely unlikely or simply impossible to happen in the analysis of 100 percent real data.

Third, the Committee studied the circumstances in which a fraud as big and long lasting as Stapel's could flourish without colleagues noticing and putting an end to it. The Committee drew several rather painful conclusions of which I mention a few. Most colleagues rather uncritically accepted Stapel's position as a brilliant and influential researcher, but the Committee also noticed that universities provided few opportunities for critical colleagues seeking support for their allegations to be effective. Stapel established, and his collegial context allowed him to establish, a working environment in which he was in control over his coworkers and PhD students, reluctant to accept criticism while keeping nosy outsiders at a safe distance. An unlikely success rate of experiments performed under his supervision and odd statistical results did not raise enough eyebrows from other scientists nor from reviewers and editors of top journals. The Committee provided a list of recommendations to universities in general, but because their focus was on Tilburg University, the recommendations were aimed at this institution in the first place. The recommendations meant to minimize the occurrence of breaches of scientific integrity such as Stapel's, and soon Dutch universities started implementing several of them and others as well. The Science Committee discussed in Chapter 3 is an example of a measure the Committee did not recommend but believed to be potentially effective.

On 28 November 2012, only 13 months after having presented their preliminary report and just a little over 15 months after the three whistleblowers revealed their findings and disclosed Stapel's fraud, the three committees chaired by Drenth, Noort, and Levelt presented their final report,[2] titled *Flawed science: The fraudulent research practices of social psychologist Diederik Stapel* (Levelt Committee, Noort Committee, Drenth Committee, 2012). They did this in a press conference in Amsterdam at the Royal Netherlands Academy of Arts and Sciences to emphasize the occasion's gravity. The Rectores Magnifici of the three universities involved also sat behind the table facing the audience and spoke words of gratitude to the committees and concern about what seemed an unprecedented breach of integrity. That same night, national TV reported on the press conference and several persons involved in the affair made their appearance in national TV talk shows.[3] The report emphasized that in addition to the identification of tainted journal articles and misled colleagues and PhD students in particular, the investigation revealed the problem of QRPs that seemed widespread beyond Stapel and the universities involved. They focused especially on the field of social psychology, which fueled some counter-reactions (e.g., Stroebe et al., 2012) but also raised the awareness in many areas of psychology and outside psychology that QRPs are a real problem that one must address. Indeed, before and after the publication of the report, several authors noticed the pervasive problem of inadequate research strategy and inadequate data analysis, presumably aimed at obtaining results that are desirable for whatever purpose the author pursues. The preliminary report published a year previously already gave away many of the main findings the final report confirmed, deepened, and enriched with more details, providing a better understanding of the range and impact of the data fraud. However, what struck many colleagues most was that Stapel used the presentation of the final report to announce on TV his testimony in the form of a book titled *Derailment*, explaining his fraudulent activities.

Apart from all the turmoil, the cold facts were that the three committees concluded that from 137 journal articles authored or coauthored by Stapel, 55 articles stemming from the Groningen and Tilburg periods were fraudulent. In another 12 articles and two book chapters stemming from the Amsterdam and Groningen periods, the committees found evidence of fraud but no proof. The oldest presumably tainted publication stemmed from 1996 and concerned Stapel's PhD thesis, 15 years before he fell from his pedestal. The reason why the committees found no ironclad proof is that data for the older publications were not available anymore. Seven PhD dissertations defended at the University of Groningen used fraudulent datasets Stapel gave his students for analysis, and for three PhD dissertations defended at Tilburg University, the Levelt Committee drew the same conclusion. One PhD dissertation, initially scheduled for defense approximately a month after Stapel's downfall, was retracted by the PhD student and never defended. The damage done to the PhD students is perhaps the most stunning of it all and may make the affair stand out from similar cases. The committees did not investigate chapters published in edited books, but the Associatie van Sociaal-Psychologische Onderzoekers (ASPO; Association of Social Psychological Researchers, which coordinates activities of Dutch social psychologists) investigated the 57 chapters Stapel contributed as an author or a coauthor to the ASPO's yearbooks (Van Dijk, Ouwerkerk, & Vliek, 2015). Although the ASPO report emphasizes that the investigation was not as thorough as that of the Drenth, Noort, and Levelt Committees, they concluded[4] that 29 publications did not meet common standards for scientific research. This brings the total number of tainted publications close to 100.

The committees found many manipulations that were not all *ad hoc* falsifications of existing data or fabrication of non-empirical data, but rather methodological trickeries to produce desired results that one would not obtain otherwise. The trickeries included the repetition of an experiment that did not produce the desired result, but in the new version introduced changes in the design, the experimental manipulation, or the materials such as questionnaires. When such design changes finally led to the desired result, Stapel reported only the successful experiment but not the initial failures, thus concealing a series of attempts that, when taken together, might have produced a desired result that was coincidental rather than a real result. This practice, where one stacks the next step on a previous undesired result without first trying to replicate that result to find support whether it is real or coincidental, readily produces a series of steps leading one in a wrong direction. This is an example of *capitalization on chance.*

Such a trial-and-error strategy may, at best, be used in the preliminary phase in which the researcher tries to get a grip on the experiment's design, but when untruthfully published as a well-thought-out one-shot experiment based on an allegedly sound theoretical starting point, the strategy qualifies as deception. A defense against chance capitalization is to repeat the experiment literally, both when the initial outcome was desired or not, and without any changes except a fresh sample of respondents, and see what happens. This is replication research, and it is common in many research areas (e.g., Open Science Collaboration, 2015). Because two instances of the same experiment may not provide convincing evidence of the truth, a series of replications is preferable, so that a trend may become visible. I will return to experiments and replication research in Chapter 5.

Other fraudulent manipulations were leaving out results of an experimental condition that did not go well with the results of the other conditions to make the total result look better. For example, a graph of the complete results may contain a dip for one of the naturally ordered conditions, but when one deletes this aberrant condition, the graph shows a regular upward or downward trend that is easy to interpret and suggests a stronger result. Reporting the smooth trend as if this is what one found in the first place and hiding the deviating condition again qualifies as deception. A third trick is to compare the results for the experimental conditions from one experiment with the control group—or benchmark group—results from another but comparable experiment when that produces the desired results. This way, one selects data from different sources and composes a hodgepodge that one presents as one experiment, again with a desired result. This is another instance of deception, and although it is different from *data* falsification and fabrication, it qualifies as FFP, but now based on *method* falsification. Closer to data fraud—falsification and fabrication—comes the combination of data from varying experiments to produce a larger sample, the exclusion or inclusion of data from respondents that make the difference in producing a desired result (e.g., the outlier example in Chapter 2), and the exclusion of subgroups of respondents because that enhances the results. Levelt et al. (2012) discussed these and other manipulations, and I advise the interested reader to consult their report. When reading, one should realize that only the fraudster's imagination limits the variety of deceptions possible.

EFFECTS OF DATA MANIPULATION ON STATISTICAL RESULTS

I return now to data and statistics. Readers not statistically versed may have difficulty understanding what changing numerical scores, duplicating real data, and making up complete datasets mean and why these practices are beneficial for someone trying to realize results at any cost and without respect for the truth. I therefore provide simple examples to clarify this and I also discuss other technical problems. These examples do not literally reflect what Stapel did but show the kinds of manipulations involved and their consequences for drawing conclusions from statistical analysis of the data. I derived much of this knowledge from speaking with other statisticians, some of whom did detective work in various fraud cases, and from using my own imagination and experience as an applied statistician for a couple of decades.

Null-Hypothesis Significance Testing

I use a toy example for the purpose of illustration. Readers needing more explanation concerning statistical null-hypothesis testing are advised to consult Appendix 4.1. Suppose I study the difference between twelve-year-old boys and girls with respect to arithmetic ability. I use an arithmetic test for measuring arithmetic ability. The test consists of 25 arithmetic problems. Students receive 1 credit point for a correct answer and 0 points for an incorrect answer. Test scores denoted X range from 0 to 25. I consider random samples of six boys and six girls; see Table 4.1 (non-bold numbers). The mean test score for boys equals 12.67 and for girls

TABLE 4.1   Artificial Data Matrix with Test Scores for Boys and Girls

Boys				Girls			
No.	X	No.	X	No.	X	No.	X
1	12	4	14	1	11	4	11
2	8/**14**	5	15/**17**	2	9	5	10/**7**
3	16/**18**	6	11	3	13	6	14/**8**

it equals 11.33, resulting in sample difference $D = 12.67 - 11.33 = 1.34$. Suppose, I want to know whether D reflects a real difference.

Small samples produce uncertain outcomes. If instead of Boy 3 with $X_3 = 16$ another boy with, say, $X = 7$ had been sampled, the boys' mean would have been 11.17, which is 0.17 units (rounded) lower than the girls' mean. A *statistical significance test* (Appendix 4.1) is used to decide whether the sample difference is large enough to reflect a real difference with high probability. Suppose I conclude that the difference of 1.34 supports a real difference; then the next question is whether the difference is small, medium, or large. I discuss the *effect size* to answer this question in Appendix 4.1. The researcher must decide whether an effect size she obtains is relevant for the application she studies. Statistics can lend a helping hand but does not carry responsibility. This is the researcher's call.

The choice of the correct statistical test depends on whether arithmetic scores follow normal distributions for boys and girls and whether the population variances in both groups are equal and known. Because it is well known and popular, I use the Student's t-test for independent samples assuming equal variances to make a decision. For this example, I did the computations by hand.[5] This may produce rounding errors (compare with Table 4.2). The data in Table 4.1 produce a value of $t \approx 0.9423$. Quantity t is a transformation of difference D and is used instead of D because it has a known distribution needed for statistical testing known as the sampling distribution (Appendix 4.1); hence, I consider the sampling distribution of t. Figure 4.1 shows the sampling distribution with mean $t_0 = 0$ corresponding to the difference between the

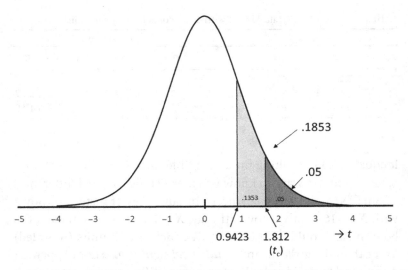

FIGURE 4.1   Distribution of Student's t-statistic for H_0, with critical t-value (t_c) corresponding with significance level $\alpha = .05$, and observed value $t = 0.9423$ (data in Table 4.1).

population mean arithmetic scores for boys and girls, and my finding $t \approx 0.9423$. The question is whether a finding is likely or unlikely when the null hypothesis of equal population means is true. To decide on this, I define a critical value, denoted t_c, such that if $t < t_c$ (in the case of a positive expectation), I say the result is consistent with H_0, whereas if $t \geq t_c$, the result is unlikely given H_0. Likeliness is defined as a probability called the significance level denoted α and often chosen equal to $\alpha = .05$. This is the area under the curve in Figure 4.1 to the right of critical value t_c.

For a one-sided t-test with significance level $\alpha = .05$ and 10 degrees of freedom,[6] tables for the t distribution provide critical value $t_c = 1.812$. Because $t < t_c$, sample difference $D = 1.34$ does not reject the null hypothesis; the probability of exceedance, p, of finding a result equal to D or greater, equals $p = .1853$, which is larger than $\alpha = .05$. The correct way to proceed is that I report there was no evidence for gender difference with respect to arithmetic ability. In scientific parlance, I report that I could not reject the null hypothesis of equal means between the two populations of twelve-year-old boys and girls in favor or the one-sided alternative hypothesis at significance level $\alpha = .05$ by means of an independent-samples Student's t-test with 10 degrees of freedom ($df = 10$) assuming equal population variances ($t(10) = 0.94$, $p = .19$).

Changing and Copying Scores

First, I change some scores to obtain a desirable result. This is a result that is significant; $p < \alpha$, which is not the case in the example. The reason significant results are desirable is that researchers are sensitive to finding "something" rather than "nothing." Given that the researcher found a non-significant result, a reprehensible decision might be that she replaces a few scores in the column for boys with higher scores; see Table 4.1, numbers in bold. This obviously will increase their group mean to 14.33. Likewise, in the column for the girls, she might replace some scores with lower scores, which has the obvious effect of lowering the group mean to 9.83. Now, $D = 4.5$. For these "data," I use the independent-sample Student's t-test again, finding $t \approx 3.1260$ with $t_c = 1.812$, rejecting the null hypothesis. The difference is statistically significant ($t(10) = 3.13$, $p = .0054$).

You will understand that research outcomes can be believable only if you accept the data as they were collected with real people, and not change the data after they were collected to manufacture a desired research outcome. Such an outcome is a hoax. One may ask what the fun is of tampering with the data and reporting a "success" that never happened. Only the fraudster knows, but it probably has to do with the prevailing culture in which significant results are highly valued and pave the road to success: articles in top journals, good reputation with colleagues, invitations to contribute lectures at conferences in beautiful resorts, access to research and travel funding, prizes, friendly faces, and the company of interesting people. To summarize, the good life, and in this respect, scientists are just like other people pursuing what they consider happiness, but some take unauthorized shortcuts.

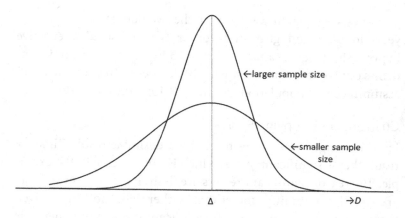

FIGURE 4.2   Sampling distributions for different sample sizes.

Second, by copying the data the subjects provided I enlarge the original dataset, so that the resulting dataset contains the same data repeatedly. Another possibility is augmenting real data with fabricated data from non-existent subjects. In both cases, adding more data lines to an existing dataset increases the sample size, and Figure 4.2 shows that sampling distributions become narrower. With a larger, real-dataset, a sample difference D that estimates a real difference Δ keeps the same order of magnitude, and if the sample grows, D eventually becomes significantly different from zero, meaning $t \geq t_c$. This happens with real data, but tainted data can mimic this result.

For the artificial data in Table 4.1, I copied the test scores to increase the sample size. This produces a highly unlikely dataset in which subjects appear as "data twins" or "data multiples," having odd effects on the outcomes of a statistical test. What happens when I copy the dataset once and add it to the existing dataset, doubling the sample size to twelve boys and twelve girls, and test again? I wish to find a significant result, so if the result of copying once is significant, I stop and report the result. However, if the result is non-significant, I copy the dataset once more, producing samples of 18 boys and 18 girls, and test again. I continue copying the dataset and testing until I find a significant result.

TABLE 4.2   Student's t-Values, Degrees of Freedom (df), and One-Sided p-Values (Independent Samples) for Datasets Consisting of Multiples of the Original Samples of Size Six

N	t	df	p	N	t	df	p
6	.9376	10	.1853	24	2.0110	46	.0251
12	1.3907	22	.0891	30	2.2581	58	.0139
18	1.7289	34	.0465	36	2.4807	70	.0078

Source: Computations were done with statistical package R.

Table 4.2 shows that the p-value is smaller as copying continues, and smaller than significance level $\alpha = .05$ when the sample sizes equal 18. One may continue to obtain any desirable result, such as rejection of the null hypothesis at $\alpha = .01$. The reason is that the sample mean difference $D = 1.34$ remains the same when I copy the complete sample results repeatedly, but the sampling distribution with standard error S_D becomes narrower. Because $t = D/S_D$, statistic t grows larger until $t \geq t_c$ (I skip some of the technical details). Provided that in the original sample boys and girls did not have the same mean test score ($D \neq 0$)—after all, why would the difference be precisely 0?—this strategy will not miss the target and surely will produce a significant result, but the enlarged datasets contain footprints of wrongdoing. That is, for boys, each of the six different scores then appears precisely 1, 2, 3, 4, 5, and 6 times producing a highly unusual distribution, and for girls four scores follow this pattern whereas score 11 appears 2, 4, 6, 8, 10, and 12 times. Real arithmetic ability scores follow a (near) bell-shaped distribution, and the sight of these unusual distributions should trigger the highest state of alarm.

Are Identical Data Lines Likely in Real Data?

Because the example sample is small, unusual distributions are easy to spot. Often, data fabrication is not this obvious, and it will be necessary to inspect the data containing the scores for all the variables for identical data lines, digging one level deeper than looking at test scores, as I did here. Perhaps the researcher copied only a few data lines or copied the data for one or a few but not all variables. Someone with a trained eye and an urge for detective work will readily write software code that identifies identical data lines. An important question is then whether identical data fragments are common in real data or whether they are rare enough to raise suspicion. I will have a closer look at the patterns of scores on the 25 arithmetic problems in a sample of generated data.

Each imaginary respondent has a generated data line consisting of 25 scores, 1s for correct answers and 0s for incorrect answers. The number of different patterns is unbelievably large, equal to $2^{25} = 33,554,432$; that is, more than 33 million. For many readers, this number may lack credibility at first sight. However, to see that this result is correct, you have to understand that each arithmetic problem has two (i.e., $2^1 = 2$) different scores, 0 and 1; two problems together have four (i.e., $2^2 = 4$) patterns of two scores, (00), (10), (01), and (11); three problems have eight (i.e., $2^3 = 8$) patterns of three scores, (000), (100), (010), (110), (001), (101), (011), and (111); and so on. Thus, the number of patterns doubles with the addition of each next item. A more vigorous example comes from the COVID-19 crisis. If we start with one infected person and each week the number of infected persons doubles, after 25 weeks we have more than 33 million infected people, and the spread of the infections will speed up if the period of doubling is shorter. You will notice the exercise is more than academic.

Is each of the 33 million data patterns likely to happen in real data? The answer is no for two reasons. First, within the groups of boys and girls, there is much variation in arithmetic ability and the 25 arithmetic problems vary in difficulty. I expect a low-proficiency student has few problems correct and that these will be predominantly the easier problems. Another student of higher proficiency has more problems correct, and these will include medium-difficulty problems in addition to the easy problems, whereas a high-proficiency student will probably only fail no more than a few of the most difficult problems. Therefore, score patterns in which students failed the easier problems while succeeding at the more difficult problems will be rare. This means that many of the possible score patterns are extremely rare. Second, many score patterns remain realistic, and their number readily transcends the sample size, rendering the occurrence of most score patterns impossible. For example, if the sample size is 500, only 500 of the millions of score patterns can be observed. Even if one cannot rule anything out completely, it is unlikely that the same pattern of 25 1s and 0s occurs twice, let alone more frequently. A small sample in which several data lines occur twice or more often should thus already raise interest. An exception is a small number of problems—remember, with three problems there are only eight different patterns, and when the sample size exceeds eight, there must be equal data lines.

To demonstrate repeated occurrence of data patterns with realistic sample sizes and, say, more than ten variables, I simulated data for 25 1/0 scored variables varying in difficulty, sample sizes ($N = 50, 100, 500, 1,000, 5,000, 10,000, 20,000$) and students varying from low to high arithmetic ability in a standard normal distribution. For each sample size, I drew 500 replications to control for peculiarities occurring in single samples. Appendix 4.2 shows technical details. The entries in Table 4.3 show proportions of patterns averaged across 500 random samples that have q copies. The first column shows results for $q = 0$ copies, that is, unique data lines. For example, for $N = 50$, a percentage equal to 99.998% of the data lines, averaged across 500 samples, has no copies, hence, occurs only once. Thus, copies occur almost never. As sample size grows to 1,000, more replicated samples contain copies, but never does a data line have more than three copies ($q = 3$), meaning it appears four times on average. However, for sample sizes up to 1,000, copies hardly ever occur. When sample size grows toward 20,000, copies appear more often, but 96% of the data lines on average are unique ($q = 0$). Results depend a lot on the choice of the person distribution and the item parameters, but it is safe to say that in smaller samples, copies are extremely rare, whereas in

TABLE 4.3 Proportion of Patterns Averaged across 500 Replicated Samples Having q Copies

N	Number Copies q					
	0	1	2	3	4	5
50	.99988	.00012	.00000	0	0	0
100	.99964	.00036	.00000	0	0	0
500	.99830	.00168	.00002	0	0	0
1,000	.99648	.00340	.00012	0	0	0
5,000	.98620	.01245	.00099	.00021	.00008	.00004
10,000	.97592	.02097	.00221	.00049	.00019	.00009
20,000	.96000	.03332	.00462	.00114	.00041	.00019

larger samples, when they occur, they occur in pairs or triplets but rarely in greater frequency.

The previous example considered 25 problems, which is not even a large number in real research, and the next computational example considers a smaller number of problems. Suppose we have four problems. The data patterns with easy problems answered correctly will happen considerably more often than other data patterns. Let the first two problems be very easy, having success probability equal to .9, hence small failure probability .1, and the other two be difficult with success probability .3, hence failure probability .7; Table 4.4 gives the distribution of the $2^4 = 16$ data patterns. Items are numbered 1, 2, 3, 4, success probabilities are denoted $P_1, \ldots, P_4$, and failure probabilities are denoted $Q_1, \ldots, Q_4$; notice that $Q = 1 - P$. By assuming fixed response probabilities, I assume these probabilities are the same for all students; hence, I assume that their arithmetic abilities are equal. This is unlikely in real life but keeps the example simple without sacrificing insight into probabilities of data patterns.[7]

Table 4.4 shows the following information. First, also in this simplified situation, several data patterns have different probabilities of occurrence, and probabilities differ greatly. Had we allowed students to vary with respect to their arithmetic ability, success and failure probabilities would have varied across students; now they vary only across problems. Hence, the probabilities of the

TABLE 4.4 Sixteen Data Patterns for Four Problems, the Computational Formulas for Their Relative Occurrence (Comput), and the Probabilities of Each Pattern (Probab).

No.	Pattern	Comput	Probab	No.	Pattern	Comput	Probab
1.	0000	$Q_1Q_2Q_3Q_4$	.0049	9.	0110	$Q_1P_2P_3Q_4$	.0189
2.	1000	$P_1Q_2Q_3Q_4$	.0441	10.	0101	$Q_1P_2Q_3P_4$	.0189
3.	0100	$Q_1P_2Q_3Q_4$	.0441	11.	0011	$Q_1Q_2P_3P_4$	.0009
4.	0010	$Q_1Q_2P_3Q_4$	.0021	12.	1110	$P_1P_2P_3Q_4$	.1701
5.	0001	$Q_1Q_2Q_3P_4$	.0021	13.	1101	$P_1P_2Q_3P_4$	.1701
6.	1100	$P_1P_2Q_3Q_4$	.3969	14.	1011	$P_1Q_2P_3P_4$	.0081
7.	1010	$P_1Q_2P_3Q_4$	.0189	15.	0111	$Q_1P_2P_3P_4$	.0081
8.	1001	$P_1Q_2Q_3P_4$	.0189	16.	1111	$P_1P_2P_3P_4$	.0729

different patterns would show a more capricious picture than that in Table 4.4. Second, in the contrived situation we study, data pattern 6, with the two easy problems correct and the two difficult problems incorrect, represents the most likely pattern with probability .3969. Pattern 11, with the two easy problems incorrect and the two difficult items correct has probability .0009, which is the smallest probability. You may notice that pattern 6 is $\frac{.3969}{.0009} = 441$ as likely as pattern 11, and in a large sample one will expect 441 as many patterns (1100) as patterns (0011).

The example shows for a small number of problems that some score patterns have a much greater probability than others do. For 25 problems, the number of possible score patterns has grown exponentially, so that even though many score patterns have probability practically equal to 0, a great many others have positive probability. Score patterns in real research follow distributions that are difficult to predict, but with larger numbers of variables—not even very large, as the 25-problem shows—the same patterns do not likely appear twice or more frequently. If one would manufacture additional score patterns, without deep knowledge of how such patterns originate, which patterns happen in practice and which patterns are highly unlikely in real data, a dataset may soon show frequencies for score patterns that are suspect.

You should have learned two things. First, it is easy to manipulate data to concoct a significant result. Second, manipulation of data produces patterns of scores that may be recognizable as unlikely to the trained and attentive researcher. So far, I have discussed changing individual scores in a dataset consisting of real data and copying complete lines of real data. When disguised as real data, both the changed scores and the copied data lines serve the purpose of producing a desired result inconsistent with the real data. Fabricating complete data lines and even a complete dataset is the same thing, and because they have the same effect on the data and the statistical analysis, in the next two sections, I will discuss both under the third kind of fraud that the Levelt Committee distinguished in their preliminary report.

THE HUMAN MIND AS A DATA GENERATING MECHANISM

A clever forger may be able to fabricate data that look real in the sense that by eyesight alone they are difficult to distinguish from real. Someone not well versed in statistics will have a difficult time convincingly forging data that are believable at first sight, because people tend to create patterns in arrays of numbers that are different from what real data show. As a result, without realizing this, they downplay the usual messiness present in real social, behavioral, and health data. In fabricated data, distributions look awkward, patterns appear in unusual frequencies, variances usually are too small, and correlations show extremely low or high values. Here are four reasons the human mind cannot properly falsify or fabricate data.

Probability. It is a well-known fact that people have trouble understanding probability (e.g., Campbell, 1974, 2002; Hand, 2014). We are used to thinking in terms of certainties—an event happens, or it does not—but probability swaps our certainty for uncertainty, and we do not like that. We like to be in control, but with probability, it looks as if we are not. Consider someone playing roulette who puts her money on red because black came out the three previous rounds, believing this reduces the chances of a fourth black outcome in a row. This is what people do, but because each spin of the wheel is an event independent of the previous spins, you might as well put your money on red all evening and have a drink in the meantime. It simply does not matter what you do.[8] Giving in to this knowledge would not only take away the meaning of playing the game—that is, the sense of being at least a little in control—but it would run counter to our belief in the causality of events. That is, we believe that we can influence events that in fact happen at random or at least predict when an event such as three times black in a row increases the probability that red turns up next time. A US roulette wheel has two zeroes on which you cannot bet and 36 nonzero numbers, 18 black and 18 red, on which you can. What is particularly demanding is to understand that winning with the first bet—with probability $\frac{18}{38} \approx .474$—or winning five times in a row—with probability $\left(\frac{18}{38}\right)^5 \approx .024$—are events based on sheer luck. There is no causality in consecutive independent spins, and you are not in control. The greatest challenge is to stop if you win in the beginning, but because many people cannot resist the fallacies of causality and being in control, in the end they will lose more than they win. Therefore, casinos can exist.

At the time of writing this text, in many countries people were concerned about possible serious side effects of vaccines against COVID-19. On its website,[9] the European Medicines Agency confirmed 25 known cases of blood clots (thromboembolic events) possibly associated with 20 million vaccinations in Great Britain and the European Union. The corresponding relative frequency was .00000125, or 1 in a million, but this nearly zero probability could not convince many people to be vaccinated with this vaccine type. Knowing that without vaccination, they ran a much higher probability of contracting COVID-19, which can cause thrombosis by the way, did not change their minds. Hand (2014) discussed rare events and showed how their occurrence misleads people all the time. A serious vaccine side effect is such a rare event but does occur now and then due to the millions of opportunities presenting themselves in a short time. The similarity with lightning striking someone comes to mind, but because thunderstorms do not occur so often in such a short time as vaccinations, we are not as frightened. All of this shows that we do not understand probability and make our decisions based on different argumentation.

Now to manufacturing data as opposed to collecting real data. Suppose you sample scores from a normal distribution; then you will have to know the probabilities—with continuous variables the cumulative probabilities—that govern the sampling process and use them. With multivariate normal data, you will also have to take the correlations between the variables into account. By construction, some variables are not normal. Think of binary data, such as gender, and discrete data, such as education type. Others follow a skew distribution or show outliers, and the selection of groups can create irregular distributions as well. When you consider all this and accept the randomness in sampling given the distributional restrictions, it is next to impossible for the human mind to produce data that look realistic. Instead, you can let a computer do the job for you, but this also sets limitations to credibility. I come back to this in the next section.

Probability is difficult to understand. Hypothesis testing also involves probabilities such as significance level, and researchers have trouble working with them properly. I once explained to a researcher that null-hypothesis testing can be considered a game of chance with "success" probability $\alpha = .05$ if the null hypothesis is true. She looked at me in dismay and said: Let us hope our research is more than that! End of conversation.

Randomness. The human mind has great trouble dealing with randomness. Kahneman (2011, pp. 114–118) presented sequences of male and female babies of different parents born in a hospital, BBBGGG, GGGGGG, and BGBBGB, and discussed people's intuition that these sequences are not randomly produced. They are, of course, because prospective mothers enter the hospital independent of one another and one birth has nothing to do with the next. If the probability of a boy or a girl being born is .5, and different births are independent, the probability of each sequence of six births equals $\left(\dfrac{1}{2}\right)^6 = \dfrac{1}{64}$. People experience the birth of children different from tossing coins and cannot resist seeing patterns believed to be inconsistent with randomness.

The tendency to see patterns people cannot believe are random events remains true for "cold" probability games entirely governed by chance, such as throwing dice. When I throw a pure die, each of the six sides has probability $\dfrac{1}{6}$ to come up. The outcome of the throw does not influence the outcome of the next throw, in which the probability again is $\dfrac{1}{6}$ for each side. Hence, the six-throw sequences (666666) and (431365) have equal probability of $\left(\dfrac{1}{6}\right)^6$, but the second sequence looks "more random" than the first. Reversely, the first sequence looks "too structured," thought impossible to be the product of chance. In fact, it is more structured than the second, considering one can summarize it as "6 6s," whereas the second sequence needs to be described verbatim, as Pinker (2021, pp. 112–113) explains, but both sequences have the same origin, which is random generation. In general, when asked to generate a sequence of numbers according to some probability mechanism, people will sooner come up with something like the second sequence. We tend to think that patterns looking structured cannot be the outcome of a probability mechanism, and (222555) and (161616) are two different examples people tend not to choose.

Digit Preference. People unconsciously have preferences for using some digits more than other digits, known as digit preference. Salsburg (2017) discussed over- and underrepresentation of digits (i.e., 0, 1, 2, …, 9) in numbers of people mentioned in religious sources, assuming that different digits had equal probability of occurrence, and found significant deviations from this expectation. He concluded that we should not take the numbers literally but have to understand them in the culture of the times in which they originated. However, in modern-day social science research, a finding like this would point to possible data fraud. Benford's law (Benford, 1938) predicts that sequences of naturally occurring numbers such as population sizes consisting of several digits—1,995,478 and 5,305,500—start with digit "1" with probability of approximately .301, digit "2" with probability ".176", and then further declining until digit "9" with probability .046.[10] For the next digit positions, the monotone decline in probability for the first digit very quickly flattens, as one would expect intuitively. Fewster (2009; see Hill, 1995, for a mathematical foundation) explained this unbelievable phenomenon, which was verified with several real-data measures.

In a social-science context, Diekmann (2007) found that approximately 1,000 unstandardized regression coefficients sampled from the *American Journal of Sociology* followed the typical Benford distribution, and in another sample, he found that the second digits followed the flatter decline in probability quite well. The author also asked students to fabricate four-digit regression coefficients that would support a given hypothesis and found that the first digits did not deviate from Benford's law but the second and later digits did. Searches for data fabrication should therefore focus in the second and later digits; deviations from Benford's law suggest faked data. It must be noted that this kind of research is still in its infancy and conclusions are premature.

Azevedo, Gonçalves, Gava, and Spinola (2021) used Benford's law to study fraud in social welfare programs and found some evidence, but Deckert, Myagkov, and Ordeshook (2011) concluded they could not use the law to describe election fraud. Such findings suggest that Benford's law might be subjected to further research concerning its empirical validity in different research areas. Without reference to Benford's law, Al-Marzouki, Evans,

Marshall, and Roberts (2005) found evidence of different degrees of digit preference (in addition to inexplicable mean and variance differences) between clinical groups allegedly selected at random and interpreted this as evidence of data fabrication or falsification. Terminal digit preference for digit 0 (and sometimes other digits) has been reported for measurements of birthweight (Edouard & Senthilselvan, 1997), blood pressure (Nietert, Wessell, Feifer, & Ornstein, 2006) and hypertension (Thavarajah, White, & Mansoor, 2003), and breast tumor diameter (Tsuruda, Hofvind, Akslen, Hoff, & Veierød, 2020), but without reference to suspicions of unwarranted data manipulation.

Variation. People tend to underestimate variation in the observed values on variables. Salsburg (2017, p. 108) noticed "lack of variability is often the hallmark of faked data." Variation is often larger, and outlying observations occur in real data, but less in faked data. Simonsohn (2013) noticed that in addition, in faked data researchers tend to create large mean differences between the different conditions in an experiment but tend to choose the within-group variances close to one another. That is, within-group variances themselves show little if any dispersion. Based on the reported means and standard deviations from an article by authors (Sanna, Chang, Miceli, & Lundberg, 2011; retracted) suspected of data fabrication, Simonsohn (2013) defined conditions that favored the reported results most and using simulated data showed that the standard deviations were highly unlikely.

The limitations on means and variances of distributions of scores are discussed in mathematical statistics (e.g., Seaman, Odell, & Young, 1985), not well accessible to researchers without statistics training. One instance that is simple and has greater notoriety concerns variables that are 0/1 scored, like the arithmetic problems discussed previously. Someone claiming her 0/1 data have variance estimate .75 is making an error or not telling the truth, because such data cannot have variance estimates greater than .5 and this only happens when the mean equals .5 and $N = 2$. In the population, let the mean be denoted μ, then the variance equals $\sigma^2 = \mu(1 - \mu)$. Thus, the mean determines the variance, and a combination of $\mu = .3$ and, say, $\sigma^2 = .15$ is impossible. A 0/1 variable follows a Bernoulli distribution, but for other variables and distributions limitations on means and variances are more complex.

Faked data are often characterized by variances that are atypically small, but large variances may also call for attention. For a variable with 100 scores running from 0 to 10 and with mean 5 the researcher reports variance estimate 25.25; what does the distribution look like? The answer is that half of the scores equal 0 and the other half equal 10, a very unlikely distribution with only extreme scores at both ends. Variances greater than 25.25 are impossible in this case, and variances not far below 25.25 are consistent with many scores at or close to 0 or 10 and few in between. Most empirical distributions are single peaked with most observations located close to the mean and fewer observations scattered in the tails of the distribution. When samples are small, in addition to a peak, distributions may show some spikes mostly caused by sampling error. Distributions may be skewed. Much depends on the design of the variable and the choice of the population. In a low-income neighborhood, education level may be heavily skewed to the right (few people are highly educated) and an easy arithmetic test may produce a score distribution skewed to the left (few students have low scores). Unexpected distributions as well as unexpected means and variances are reason for further scrutiny.

The final example concerns someone who reports that for a normal variable 60% of the scores lay between -1.96σ and 1.96σ from the mean. This person is making an error or not telling the truth; the right answer is 95%. Perhaps the distribution was not precisely normal. In principle, any distribution is possible, and each distribution has a mean and a standard deviation, but the opposite is not true. That is, a distribution's mean μ and standard deviation σ limit the shape of the distribution of the scores; see the example in the previous paragraph for 0/1 variables. Chebyshev's inequality[11] informs us that for any distribution with known mean μ and standard deviation σ, at least 75% of the scores must lie between -2σ and 2σ (rather than using 1.96, which is typical of the normal distribution). Someone claiming that 60% of the scores were in this interval is making an error or not telling the truth.

The previous discussion revealed an interesting distinction between faking data and faking results. Results are, for example, the summaries of the data resulting from the application of statistical methods, such as a significance test or a regression analysis producing a regression model. A fabricated dataset has means and variances and probably other features as well that are consistent with the data; that is, not the results are fabricated but the data are, which does not render the results useful, of course. Faked results that do not originate from underlying data, faked or real, can sometimes be recognized when in combination they look unusual or even seem impossible. Such recognition should lead to requests to see the data and emphasize the need for having data posted at an easily accessible repository.

Researchers interested in checking or revealing data falsification or fabrication are looking for these and other features, but also for values of test statistics, degrees of freedom, probabilities of exceedance, and effect sizes that seem incompatible. Effect sizes may be unusually large compared to what is common in a research area. This kind of detective work proves quite difficult unless the fraud was drastic, and evidence piles up so high one cannot ignore it. I am not implying one should not pursue this avenue—being a statistician, I fully support the further development of a methodology of detecting statistical irregularities—but from a legal viewpoint, nothing beats a confession from the alleged fraudster. At Tilburg University, we were particularly lucky that Stapel, when confronted with the allegations brought forward by three whistleblowers, confessed to the fraud, and even cooperated in identifying an initial set of tainted articles—until he decided it was enough. Many fraud cases linger on for a long time because it proves difficult to come up with the statistical evidence and the alleged fraudster does not cooperate; of course, it is everyone's right not to implicate oneself.

THE STATISTICAL MODEL AS A DATA GENERATING MECHANISM

The human mind is a notoriously bad information-processing machine, as many psychologists have documented (e.g., Arkes, 2008; Grove & Meehl, 1996; Kahneman, 2011; Kahneman & Tversky, 1982; Meehl, 1954; Tversky & Kahneman, 1971, 1973, 1974; Wiggins, 1973). The generation of a believable dataset is an incredibly demanding task for the human mind. Knowing this, it is the more remarkable that some people try to accomplish this and think they can get away with it. Fraudsters not trusting their ability to forge data "by hand" may resort to data simulation, much as I did in this book. However, this requires quite some knowledge and understanding of statistics, not obvious from the data fraud cases I know. I am unaware of anyone who ever followed the route of data simulation using statistical models, and I will not recommend it. Statisticians use simulated data when they study properties of statistical tests and distributions, and the math becomes too complicated so that ironclad mathematical derivations or proofs become awkward. They also resort to simulated data when they compare different statistical methods designed for the same purposes to find out which one is best.

Data simulation uses a particular statistical model and particular distributions of variables; in Appendix 4.2, I use a model for response probabilities and assume that proficiency follows a standard normal distribution. This way, I control the population distribution, the sampling model, and the way in which a probability mechanism produces answers to problems depending on features of the people and the problems. I am also able to include violations of assumptions, although irrelevant here, to study how these affect statistical tests, but the crucial difference with a fraudster's intentions is that the statistician reports doing this and explains how and why. For statisticians, simulating data is a sound strategy to learn how statistical methods work in an environment they can control so that they know how the methods react to population and sample data features. Simulated data, however, when presented as real data, are fully inappropriate for pretending you are studying real phenomena, such as motives of people disobeying public regulations during a pandemic or the physical reactions of people to an experimental vaccine against the COVID-19 virus. The results are worthless.

For the moment setting ethics aside, you could use this strategy to fabricate data and pretend they come from real people. Like a statistician in control of the situation that enables her to assess a statistic's performance, the researcher is now in control of the desired power (Appendix 4.1) of the statistical test and the effect size. However, just as *handmade* data may produce a result that is too good to be true, one can say the same of *simulated* data presented as real data. Whereas the handmade data often contain unlikely features reflecting researchers' inability to act as credible data sources, the model-simulated data are too smooth to pass for real. The model I used for simulating data typical of students solving arithmetic problems is an idealization of how this might work, assuming a correct answer depends on the proficiency level and two problem features, which are how well the problem distinguishes low and high-proficiency students and at which difficulty scale level it does this. Social, behavioral, and health researchers use such models to summarize important real-data features and to help decide whether the number of correct answers to the set of problems is justified as a performance summary. Anyone understanding what a model is and having worked with models for data analysis, knows that models never describe data perfectly; by design, they are too simple for that purpose and only single out important features but lose information from the data in the process. This is what a good model does: It emphasizes important features and ignores the rest, and by doing this, at best approximates the data.

Real data are much messier than a model suggests, and next to signals, many of which are unexpected and unintelligible, real data contain much noise. Without going into detail, real data probably contain information about the researcher's hypothesis, but the information often is less pronounced than one hopes, and much of the variation in the data remains a mystery to the researcher. How this messiness can invite an exploratory attitude with uncertain outcomes is a topic of Chapter 5. A researcher reporting a model that fits her data except for sampling fluctuation should expect at least some professional reservation of colleagues and critical debate, for example, instigating other researchers to re-analyze the data. This means that data must be available to other researchers interested in a re-analysis, a situation that still is not default in large parts of the scientific community (Chapter 7). Debate and data re-analysis may lead to the conclusion that the work is in order but given the abstractness of models relative to the messy reality of data, one may anticipate some criticisms of the work.

My point is that anyone contemplating simulating data from a statistical model with the intention to present results based on them as if they originated from real data may find the enterprise will fail. Knowing this disclaimer on direct use of simulated data as if they were real, one might extend the model with inexplicable signals and random noise. However, making models more complex does not free them of certain smoothness features in the simulated data. In addition, the simulated data may show traces of the artificiality of handmade data, due to having to design the additional model features. The unexpectedness and incomprehensibility of real data features are hard to mimic by the human mind. However, I cannot know whether data generation using statistical models has not already taken place, escaped colleagues' attention, and is a next phase in an "arms race" between fraudsters and the scientific community. We must expect data to be noisy and beset with unintelligible signals and show a healthy skepticism toward perfectly fitting models, requiring at least a second check. During the Stapel affair, one of my PhD students stepped into my office disappointed, explaining that the analysis of his data had produced results that looked messy and inconsistent with expectation. I tried to cheer him up saying that at least we knew for sure that his data were real. Fortunately, he could take the joke.

FINAL REMARKS

Recognition of falsified or fabricated data (or results) is more likely when colleagues are keen at discovery, but usually they are not, and the swindle remains unnoticed. The freely available software program statcheck[12] (Epskamp & Nuijten, 2016) scans documents such as articles for p-values that are inconsistent with other reported statistical results and signals such inconsistencies. The program is a valuable tool for finding signs of possible data fraud, but inconsistencies may also result from (unintentional) QRPs and accidents (Chapter 2). Nuijten, Hartgerink, Van Assen, Epskamp, and Wicherts (2016) found evidence of a systematic bias in literature reporting inconsistent p-values toward researchers' expectations. Researchers keen on avoiding errors in their own work may also use the program to correct reporting errors and protect themselves from QRPs.

I have two different remarks. First, when they are motivated, data fraud detectives must exercise the same methodological rigor as anyone doing scientific research. Given the messiness of data, studying a dataset for irregularities will almost surely produce results, also when everything is in order. Finding a suspect result that is not the result of fraud is a false positive. Exploration, one of the two topics of the next chapter, lurks in each study, and detective work is no exception. The data fraud detective must always be aware of the risk of taking a phenomenon in the data seriously when other explanations are plausible as well. Some authors (e.g., Klaassen, 2018) have argued that in science the leading principle must be *in dubio pro scientia*—when in doubt, decide in favor of science—meaning that the published article under suspicion should be retracted, even when the evidence is not ironclad. The flipside of this viewpoint is that it puts enormous strain on the data fraud detective. She should not only know she has to convince the defendant—who clearly has other interests—but also colleagues who may not be impressed by the evidence presented. Moreover, retraction of an article not only raises negative attention with colleagues and sometimes the media, but also damages the researcher's reputation. In that case, let the evidence be impeccable.

Second, when colleagues do not pick up possible data fabrication or falsification, this does not mean they are indifferent to what their fellow colleagues do. In my opinion, this apparent inattentiveness reflects that people act and think based on trust. They simply and tacitly assume that other people can be trusted to do the right things, even if they do not do them flawlessly all the time. No one is without error, and people look at themselves when rumor at the water cooler has it that a trusted colleague may be in error. It takes time, experience, an independent position, some reputation, and courage to raise a hand and make an accusation that may well backfire. Adopting mistrust of others as a basic attitude beyond a constructively critical and enquiring approach would probably make it very hard to function well in a scientific environment, let alone daily life. However, introducing research policy that requires everybody to perform her research in a transparent way may be more effective while maintaining a livable work atmosphere. I already mentioned publishing one's data as an institutional policy and add involving a statistician in one's research team to do the work they master better than most researchers do. Prevention seems more effective than picking up the pieces. This is the topic of the final Chapter 7.

TAKE-HOME MESSAGES

The human mind has great trouble understanding probability, randomness, digit preference, and variance. These are conditions complicating the believable fabrication and falsification of data.

Statistical models can be used to generate sample data that often are too smooth to be true. Recognizing artificiality is difficult, calling for maximum transparency of all aspects of research.

Fabricated or falsified data have features unlike real data that a trained eye can often recognize without legal proof of malpractice. A fraudster's confession is worth more than statistical evidence of malpractice.

APPENDIX 4.1: STATISTICAL TESTING, POWER, AND EFFECT SIZE

In this appendix, I discuss statistical concepts concerning null hypothesis testing. I use the arithmetic example comparing boys and girls discussed in this chapter and the Student's t-distribution. This appendix is read best in combination with the example discussed in this chapter.

Statistical testing works as follows. Consider the populations of boys and girls. Population mean μ_{BX} (μ: Greek lower-case mu) denotes boys' mean arithmetic score, and μ_{GX} denotes the mean for girls. The difference $\Delta = \mu_{BX} - \mu_{GX}$ (Δ: Greek capital delta) is of interest. First, consider the case that boys and girls do not differ. The null hypothesis H_0 representing this case is,

$$H_0 : \Delta = \mu_{BX} - \mu_{GX} = 0$$

Second, for reasons of simplicity I will discuss a one-sided rather than a two-sided test. I assume that the researcher wants to know whether boys on average have higher arithmetic test scores than girls; then the alternative hypothesis H_A is

$$H_A : \Delta = \mu_{BX} - \mu_{GX} > 0$$

The null hypothesis is the benchmark in significance testing. Suppose it is true. As a thought experiment, I repeatedly draw random samples containing six boys and six girls. For the same gender group, samples differ by coincidence, called sampling fluctuation. For each pair of samples, I compute the means, $\bar{X}_B$ and $\bar{X}_G$, and their difference $D = \bar{X}_B - \bar{X}_G$. Because means vary across samples, so does their difference D. Under the null hypothesis, the mean of D equals $\Delta = 0$. Figure 4.2 (main text) shows two sampling distributions for D, the widest for a small sample size, thus reflecting that small samples provide less certainty about population difference Δ. Larger samples produce narrower sampling distributions, hence provide more certainty about the population difference Δ.

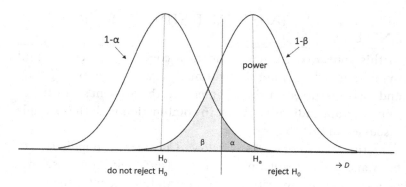

FIGURE 4.3 Sampling distributions under the null and alternative hypotheses.

Next, assume the alternative hypothesis is true; then repeated sampling produces a sampling distribution with mean $\Delta > 0$. For a fixed sample size, Figure 4.3 provides the sampling distributions for the null hypothesis of no difference (left) and for the alternative distribution of positive difference (right). The logic of null-hypothesis statistical testing is assessing whether the data provide evidence to reject the null hypothesis in favor of the alternative hypothesis. How does this work?

In the example, I found that the mean arithmetic score for boys is 1.34 units greater than the mean score for girls (Table 4.1, main text). I discussed statistical testing with a significance level α, in research often $\alpha = .05$; the corresponding critical value of the test statistic, t_c in case of a Student's t-test; and a probability of exceedance, p. Figure 4.1 (main text) shows these quantities. Significance level α is the area under the null distribution of statistic t to the right of critical value t_c, and thus is a probability, here, of finding values t greater than t_c; that is, located in the critical area, $t \geq t_c$. When the researcher choses $\alpha = .05$, under the truth of the null hypothesis, she has a .05 probability of finding a result greater than the critical value. If $t \geq t_c$, then the null hypothesis is rejected.

Because this would be an incorrect conclusion, probability α is also called the Type I error probability. Researchers more reluctant to reject the null hypothesis chose a stricter significance level, such as $\alpha = .01$, $\alpha = .005$, or $\alpha = .001$, but sometimes someone choses the more liberal $\alpha = .10$.

A result demonstrated in this chapter but difficult to grasp is that any *fixed* non-zero value of a sample statistic, such as D, becomes significantly different from zero if the sample size continues growing long enough. Researchers sometimes are happy with a significant result, but the question is whether the effect is worthwhile. For example, is a difference of 1.34 on a particular arithmetic test on a scale running from 0 to 25 small, medium, or large for decisions one wishes to make about twelve-year-old boys and girls? The *effect size* is used to qualify a particular effect as small, medium, or large, or to use other qualifiers for identifying sizes.

Figure 4.3 shows significance level α and three additional probabilities. Whereas α is the probability under the null hypothesis of finding a t-value in the critical area, $1 - \alpha$ is the probability of finding a result consistent with the null hypothesis. The sampling distribution of t if the alternative hypothesis H_A is true also shows two probabilities. Then, β is the probability of $t \leq t_c$, meaning not rejecting H_0, which is false; that is, incorrectly not rejecting H_0. Probability β is also called the Type II error probability. Finally, $1 - \beta$ is the probability of correctly rejecting H_0 in favor of H_A; this is the *power* of the statistical test. The power is extremely important. Based on the minimum effect researchers wish to establish with enough certainty, researchers determine the corresponding sample size minimally required to reject the null hypothesis in favor of the alternative hypothesis. This works as follows.

First, if the alternative hypothesis is true, due to a larger power, a larger sample makes it easier to reject the null hypothesis for a given difference Δ and the corresponding difference t_Δ.

Second, the researcher choses a value for Δ under the alternative hypothesis, such as Δ = 3, and a value for the desired power 1 − β of identifying this value when, say, α = .05 under the null hypothesis of no difference, Δ = 0. Let us assume that she choses 1 − β = .7; that is, the probability of rejecting the null hypothesis when Δ = 3 equals .7. The reasoning is: "When I use α = .05, I want to be able to find a true difference at least equal to Δ = 3 with probability 1 − β = .7. What is the minimum *sample size* I need to accomplish this?" By a kind of backward reasoning that I do not discuss here, one can compute the minimum sample size needed.

A final word is that there are several ways to assess the meaning of a research result. Bayesian statistics provides an alternative (Chapter 7), but a simpler approach is to estimate the result's *standardized effect size* (Cohen, 1960; 1988, p. 20). In our example, this would be the value of D one finds in one's data, corrected for the observed range of arithmetic scores in the groups of boys and girls and only reported when D is significant. Suppose for larger samples than I used in the examples, $D = 1.34$ would have been significant at α = .05. What does this mean for the magnitude of the difference in addition to knowing it is not zero? Suppose in the two larger samples of boys and girls the arithmetic scores are spread across the whole scale from 0 to 25; see Figure 4.4, upper panel, where the score distributions are shown as continuous for convenience, thus ignoring that the arithmetic score only has discrete values. Then, a mean difference of 1.34 is relatively small. However, when the arithmetic scores within groups are much closer as in Figure 4.4, lower panel, difference 1.34 is relatively large. Whether a difference is important depends on the use of the difference made by, for example, policy makers. The effect size is the difference corrected for the standard deviation of the differences in the sample, $d = \dfrac{D}{S}$; notation S denotes the standard deviation in the sample. I refrain from further computational details.

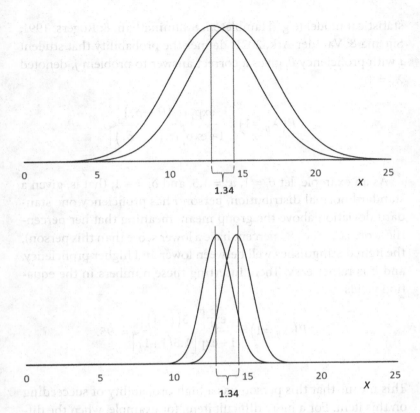

FIGURE 4.4   Two wide, largely overlapping distributions (upper panel) and two narrow, more separated distributions (lower panel), both with differences between means equal to 1.34.

APPENDIX 4.2: SIMULATING DATA

For the simulated dataset, I assumed a test with 25 problems that are indexed $j = 1, ..., 25$. Problems varied to the degree they could distinguish students with low proficiency from students with high proficiency (discrimination parameter α_j; index j shows that the value of the parameter depends on the item) and had varying difficulty levels (difficulty parameter δ_j). Parameter θ (Greek lower-case theta) denoted students' proficiency at arithmetic and followed a standard normal distribution, $\mathcal{N}(0,1)$. The next statistical model (e.g., Hambleton, Swaminathan, & Rogers, 1991; Sijtsma & Van der Ark, 2021) defines the probability that student i with proficiency θ_i gives a correct answer to problem j, denoted $X_{ij} = 1$:

$$P\left(X_{ij} = 1\right) = \frac{\exp\left[\alpha_j\left(\theta_i - \delta_j\right)\right]}{1 + \exp\left[\alpha_j\left(\theta_i - \delta_j\right)\right]}.$$

As an example, let $\theta_i = 1$, $\alpha_j = 1.5$, and $\delta_j = -1$; that is, given a standard normal distribution, person i has proficiency one standard deviation above the group mean, meaning that her percentile score is 84 (i.e., 84 percent have a lower score than this person), the item distinguishes well between lower and higher proficiency, and it is rather easy. Then, inserting these numbers in the equation yields

$$P\left(X_{ij} = 1\right) = \frac{\exp\left[1.5\left(1+1\right)\right]}{1 + \exp\left[1.5\left(1+1\right)\right]} \approx .95.$$

This means that this person has a high probability of succeeding on this item. For a more difficult item, for example, when the difficulty matches the person's proficiency level (i.e., $\delta_j = \theta_i = 1$), one finds $P(X_{ij} = 1) = .5$. The reader can check that difficulties greater than 1 produce success probabilities smaller than .5. Randomly drawing numbers with equal probability from the interval [0, 1] and comparing them to the probability of interest transforms the latter to 1/0 scores. For example, if the response probability is .8, the random draw from [0, 1] has a probability of .8 to produce a number between 0 and .8, and produces problem score 1; otherwise, the score equals 0 with probability .2.

For the example, I drew samples of person parameters θ at random from the standard normal distribution. Table 4.5 shows the discrimination and difficulty parameters.

TABLE 4.5 Discrimination and Location Parameters for 25 Arithmetic Problems

Item	Discr	Diff	Item	Discr	Diff
1	1.0074782	0.826587656	14	1.0959253	0.741778038
2	0.8067685	1.353150661	15	0.8580500	0.002012919
3	0.9269077	-1.041643475	16	0.9288094	-0.898064550
4	1.1931021	1.083086133	17	0.5519033	1.084384221
5	0.5851360	0.576409022	18	0.7641777	-1.942264376
6	0.7254366	0.142388151	19	0.8987907	0.915858250
7	0.7745305	-1.627647482	20	1.3361341	-1.000478113
8	0.7723051	-1.320787834	21	1.3647212	-1.355266893
9	1.1158293	1.599329803	22	1.1153524	-1.931829399
10	0.9296715	-0.309449572	23	1.2751099	-0.055598619
11	1.1516557	0.990985871	24	0.8555687	-1.588399312
12	1.0677378	1.290610320	25	0.9058500	1.206188018
13	0.6135090	1.818614585			

NOTES

1 https://ktwop.files.wordpress.com/2011/10/stapel-interim-rapport. pdf

2 https://www.rug.nl/about-ug/latest-news/news/archief2012/ nieuwsberichten/stapel-eindrapport-eng.pdf

3 https://www.bnnvara.nl/dewerelddraaitdoor/videos/226619; interview with Ed Noort, chair of the Noort Committee, University of Groningen, in Dutch. https://www.npostart.nl/pauw-witteman/28-11-2012/ VARA_101293877; interview with Pim Levelt, chair of the Levelt Committee, Tilburg University, in Dutch.

4 http://www.sociale-psychologie.nl/wpcontent/uploads/2015/06/ Verslag-ASPO-commissie-van-Dijk.pdf

5 Unbiased estimates for variances of X were 8.6667 for boys and 3.4667 for girls. The pooled variance was 6.0667. Based on the pooled variance, variances of sample means were 1.0111 for both boys and girls. The variance of the difference of sample means was 2.0222 and the standard error was $S_D = 1.4221$. The t-value was $t = D/S_D = 1.34/1.4221 = 0.9423$.

6 In the computation of a test statistic, such as the Student's t-test, what matters is the maximum number of data points that can vary freely, hence, are not fixed when other data points are known. For example, in a test statistic that uses the mean of, say, six sampled scores on a variable, knowing this mean and five of the six scores determines the sixth score. The number of degrees of freedom is 5 whereas the sample size is 6.

7 I assume that in computing the probability of a particular pattern, I can multiply the success and failure probabilities, thus ignoring the possibility that trying to solve one problem can have an influence on the probability of solving consecutive problems. Hence, I assume that each problem stands on its own, and when trying new problems, one does not benefit from having tried previous problems. This is a common assumption, known as conditional independence, in many statistical models that avoids models becoming too complex.

8 For a US roulette wheel, there are 2 zeroes (not allowed for betting; the casino's insurance policy) and 36 nonzero numbers, 18 black and 18 red. If you bet on red all the time, your probability of winning is $\frac{18}{38}$ and of losing $\frac{20}{38}$, and the same probabilities hold when you bet consistently black. Your *expected* yield for each dollar is

$$E(\$) = \frac{18}{38} \times 1 + \frac{20}{38} \times -1 = -\frac{2}{38} \approx \$ -0.0526. \text{ If you would bet at}$$

random, red and black have the same probability of turning up, which is $\frac{18}{38}$, and losing in either case has probability $\frac{20}{38}$; hence, nothing changes.

9 Downloaded on 19 April, 2021, from https://www.ema.europa.eu/en/news/covid-19-vaccine-astrazeneca-benefits-still-outweigh-risks-despite-possible-link-rare-blood-clots

10 The equation for the first digit is the following. Let $P(d_1)$ be the probability of d_1, $d_1 = 1, 2, ..., 9$, ignoring leading zeroes, and let $\log_{10} X$ denote log base 10 of X, then $P(d_1) = \log_{10}\left(1 + \frac{1}{d_1}\right)$ gives the desired probabilities.

11 Chebyshev's inequality (Freund, 1973, 1993, pp. 217–219) sets a lower bound on the proportion of scores on random variable μ between $-k\sigma$ and $k\sigma$, by $P(-k\sigma \leq X - \mu \leq k\sigma) \geq 1 - \frac{1}{k^2}$. For two standard deviations above or below the mean, one finds that at least a proportion $1 - \frac{1}{2^2} = .75$ of the observations are between these two bounds.

12 Downloadable from https://mbnuijten.com/statcheck/

CHAPTER 5

Confirmation and Exploration

T HE DISTINCTION BETWEEN CONFIRMATORY and exploratory research appeared in some places so far and is essential for understanding several of the QRPs. This is also the distinction between theory-driven and data-driven research, between testing an existing theory and searching the data for interesting features, and between working according to plan and without much of a plan and letting the data speak. The multiple regression example in Chapter 2 illustrated exploration. In this chapter, I will argue that theory-driven research is the superior of the two strategies, and that data-driven research is useful when it leads to a new theory or amends an existing theory. When misrepresented as confirmatory, publications based on exploratory research convey the impression that the authors are more certain about their results than the truth allows. In this chapter, I will warn against presenting unplanned research as if the researcher planned it and explain the risk of misrepresenting the research's status.

Because it is goal-oriented and explicit, and because it helps construct a new theory or produces new knowledge about an

DOI: 10.1201/9781003256847-5

existing theory, many colleagues will acknowledge the greater value of confirmatory research. The downside is its vulnerability due to betting on one horse. Losing—for example, not finding a result one expected—at first sight means staying behind empty-handed, but the upside is that one learns whether one's theory is correct and if not, how one should amend it and put it up to the next test. This rather harsh, black-and-white scenario, which develops in small steps forward with an occasional relapse, thus requiring patience and resilience, is not attractive to many researchers. This is certainly true in the social, behavioral, and health sciences that usually do not produce theory that passes the test when empirically examined. Moreover, influences researchers find hard if not impossible to control confound the data, rendering success in theory testing, even if within reach, a difficult enterprise. Therefore, it is perfectly understandable that the prospect of confirmatory research puts off many researchers, and that they prefer the venue of exploration that allows them multiple researchers' degrees of freedom; that is, making decisions on how to move forward *during* the data analysis rather than *prior* to it. This is precisely where the danger lurks, and even in experimental research, which is explicitly confirmatory and directed at hypothesis testing, researchers find it hard to accept an unanticipated, often disappointing outcome and sometimes tend to change course during the data analysis and revert into the exploratory mode rather than returning to the drawing board.

All of this would not be problematic, at least not dramatically so, when research reports would present precise accounts of the theory, the hypotheses, the research design, and the data analysis. This way, reviewers of preliminary publications would be able to comment on the methodological and statistical quality of the research and single out the ones that try to sell chalk for cheese. Such detailed accounts are usually absent because researchers do not keep them or because journals do not appreciate publishing logs of what researchers did precisely before arriving at a result. The point of this chapter is that in not recognizing the exploratory nature of the research, readers take the results, many of which would fail upon replication, too seriously. This puts them on the wrong track for their own research, not knowing that it has a high probability of failure, thus wasting precious time and money. In addition to preferring the friendlier exploratory research strategy to the tough world of confirmation and falsification, I will touch upon problems caused by academic policies that reward researchers that show results with a positive message.

Before I do, I anticipate a possible critical reaction that I might be against exploration whatsoever: I am not. Regularly, researchers come across interesting results in their research they did not expect, and which are too compelling to ignore. My position is simply that they should follow the lead suggested by an unexpected and interesting result and report it as post hoc or exploratory, preferably suggesting ways to follow up on the result using a confirmatory research design.

CONFIRMATION AND EXPLORATION AS FISHING EXPEDITIONS

The problem that exploratory research causes when presented as confirmatory without all the decisions taken between data collection and publication is quite difficult. How can one begin to understand the problem? Let us look at a metaphor that I hope illuminates it. I use fishing, perhaps not popular among many readers, but used by critics (e.g., Wagenmakers, Wetzels, Borsboom, & Van der Maas, 2011; Wagenmakers, Wetzels, Borsboom, Van der Maas, & Kievit, 2012) of exploratory research when they refer to exploration as fishing expeditions. This is what I think they mean.

Suppose you set out fishing but have no knowledge of it. Here are two strategies. The first strategy is that you study all there is to know about fishing by reading about it and talking to experienced anglers for practical leads. Then, you decide about the kind of fish you would like to catch, buy the equipment, and choose the right time of the year and the most likely location for catching the fish. When the day has come, you may or may not find that you are successful and in either case, you may decide to try a couple of additional days to find out whether the result holds upon repetition. Different days may show different results, but no matter the outcomes, the prolonged effort will teach you a lot about fishing. For example, you may conclude whether you were lucky the first day and that the fish are at that location only occasionally. You may also conclude that you can do better and think about improving your skills. After some trials, you may find that your improved strategy guides you to consistent results.

The other strategy takes less time for preparation, has more room for trial and error, but may nevertheless produce results. You skip the thorough preparation and right away buy a huge net and drive to the nearest lake, pull the net through the water, and see whether something comes up and whether that includes the target fish. It may, but chances are that your fish is not among the catch, because you prepared less well this time and guessed. The net may nevertheless contain other kinds of fish and lots of waste like empty cans and plastics. You decide to shift your interest from the target fish to other kinds that happen to be there. Does this practice improve your knowledge about fishing? Not much, but when you come home and bring the other fish, at least you are not empty handed and your acquaintances might even applaud at your sense of entrepreneurship, your smooth shift from one goal to the next.

In daily life, taking a long shot, not sticking to one goal when that appeared difficult to realize and shifting to another goal that presented itself, sometimes is a clever strategy, but this flexibility can be problematic in science. The first strategy's strength is that, even if it took a lot of time and energy, it uses a well-thought-out plan, and when flawed, it often helps you take one or two steps ahead in understanding the problem. Applied to fishing, accumulating knowledge rejecting some steps and accepting others based on evidence may develop your skills and lead to you to fertile fishing grounds. The second strategy provides you with few if any cues of the causes of the outcome, and learning opportunities are rarer. However, you have a result, but the problem is that the result was unexpected and not based on prior expectation founded on careful preparation and design; *it simply happened, and you do not know why.* You do not even know whether you were lucky. You have no way of knowing this without some prior knowledge to guide your thoughts. Repetition checking whether results change or stay the same can provide you with more knowledge, but this assumes you shift to the first strategy after all.

Researchers using the second strategy may not be aware that it allows coincidence to exert a greater role on the research outcomes, and that one study cannot tell what is really going on. Especially in large datasets, there will always be significant results, interesting effect sizes, and correlations, simply because data are never completely random, containing only noise and no signals. Which force would have created such datasets consisting only of noise unless they did it deliberately using a computer for simulating randomness? Moreover, even if data consisted of only noise, on average, one out of every 20 tests would be significant—assuming significance level $\alpha = .05$—and interesting effect sizes would appear simply by chance.

Table 5.1 shows results of data I simulated according to a model that represents coin flips to produce scores 1 (head) or 0 (tail). In the context of the arithmetic study discussed in Chapter 4, this choice would mean that the probability of having any of the 25 problems correct equals $P = .5$, irrespective of students' proficiency. When you think of it, this means that I somehow managed to select a sample of students that all have the same proficiency level (how did I know without the test?) and designed a set of arithmetic problems that all are equally difficult (they must resemble one another a lot), a most unlikely situation. However, my goal is to demonstrate something else, which is that even with coin flipping data, one will find results that may look interesting without further scrutiny.

I chose samples of 50, 100, and 500 students; that is, as many as 50, 100, and 500 coin flips in each of the 25 columns of the data.

TABLE 5.1   Sample Size, Results Averaged across 100 Random Samples of Size N. Ranges of Minimum and Maximum Sample Correlations, Minimum, Maximum and Mean Numbers of Significant Results. Critical Values for Null Hypothesis Testing

N	Range Min	Range Max	Min (Sign)	Max (Sign)	Mean (Sign)	Crit Value Corr
50	-.60; -.29	.30; .64	6	25	15.32	.28
100	-.42; -.21	.21; .40	7	23	15.34	.20
500	-.18; -.09	.10; .18	8	28	15.60	.09

I know that coin flips produced the data, so I expect that all 300 unique correlations[1] between problems are 0 (300 null hypotheses H_0: $\rho = 0$ against alternative H_A: $\rho \neq 0$;[2] Greek lower-case rho for correlation in population). Given significance level $\alpha = .05$ and 300 independent null hypotheses, I expect $\alpha \times N = .05 \times 300 = 15$ significant results[3]. I replicated each sample 100 times to control for accidental results caused by random sampling. Results in Table 5.1 provide averages across replications.

Especially when samples are small or modest, sample correlations can be interesting judging their magnitude. For example, for $N = 50$, the smallest correlation found in 100 datasets was −.60 and the largest was .64. These values make interesting "effect sizes" when one is looking for results without a theoretical point of departure. In this example, effect sizes diminish as sample size grows. Mean numbers of significant results are approximately 15, as expected, irrespective of sample size. I emphasize once again that the data result from coin flipping and are meaningless. The human eye may judge differently, of course, but nevertheless incorrectly.

There are statistical protections reducing the probability of finding anomalies and so-called multivariate methods to test our hypotheses reducing the risk of making incorrect judgments (e.g., Benjamini & Hochberg, 1995; Maxwell & Delaney, 2004; Pituch & Stevens, 2016; but see Gelman & Hill, 2007, for a different perspective). However, because I wanted to demonstrate that even with coin flips, you may find "something," I refrained from using correction methods. You should memorize two things. One is that even coin-flip data can contain results other than zero correlations. The other is that real data are not the result of coin flips or another mechanism that produces a neat data landscape, as flat and barren as a desert with a tree sticking out here and there that is a fata morgana on closer inspection. Real data are much bumpier, and there is much more to see, suggesting results that look interesting but need not be interesting. Data always contain a lot of structure. I will have a look at this phenomenon later in this chapter.

The fact that there is always something to find in datasets emphasizes the need for a plan when doing research. Without it and a design based on it, the researcher does not know what she sees. The need for publishable results and our talent to assign meaning to almost anything we see lures us into reading volumes into significant results, effect sizes, and correlations we never expected (e.g., Kerr, 1998). Most of these results will not stand the test of replication (Ioannidis, 2005). When you have an idea where to look and design the conditions for discovery, this may at least put you on a trail that may lead you somewhere. If the great scientific discoveries would have been there for the taking, their unfolding would have been so much easier, but scientific progress is a hurdle race where many participants never finish.

CONFIRMATORY AND EXPLORATORY EXPERIMENTS

Let us have a look at real scientific research and consider the confirmatory version and the exploratory version. Rather than presenting two versions of the same project as if researchers can freely choose one, I will discuss the two versions as parts of one study that typically starts confirmatory and then transforms to exploration. I will discuss two projects that I made up so as not to embarrass anyone. First, I will discuss an experiment and then a survey, and discuss for both types what happens when one gears from confirmation to exploration. In addition to discussing the research question and the way the researcher executes the research, I will also explain the logic of an experiment and a survey.

I consider a simple experiment intended to demonstrate that a new drug, which I label X, has a negative effect on people's ability to drive their car safely, which I label Y. Thus, the question is whether X influences Y. The researcher assesses driving ability in a traffic simulator for obvious reasons. Now, if the researcher would provide one person, say John, only one dose of the drug, say, 10 mg, she would observe a particular level of driving ability for John, but this would not tell her anything about X influencing Y. If John shows adequate driving ability, the researcher cannot conclude that the drug is *ineffective* because she does not know how John reacts when given another, even a lower dose. In addition, if John shows inadequate driving ability, the researcher does not know whether this is simply his driving style irrespective of the drug. This means that the researcher needs to distinguish different conditions in her experiment, each with a different dose. The condition with the zero dose represents a baseline for driving ability, called the control condition. A participant might react to knowing the dose they received, even if the drug does not affect the participant physically. Thus, the participant must not know the dose. Then, we call the experiment blind, and it is double blind if also the researcher does not know before and during the experiment who received which dose (and someone else keeps the administration). This way, it is impossible for the researcher to provide cues accidentally to the subject to which she might react, and which might affect Y. In the control condition, a participant will receive a placebo, which looks like the drug but does not contain an active ingredient, without knowing it is placebo.

Experience has shown that different participants tend to react differently to the same treatment. Therefore, each condition contains a random sample of participants from the same well-defined population. The inclusion of multiple participants in each condition allows the assessment of the variation in participants' reaction to the dose given in the same condition. In addition, the researcher can assess whether on average participants in different conditions react differently to different dosages of the drug. Statistically, this comparison of conditions entails estimating in each condition the mean and the variance in driving ability. The F-test (e.g., Hays, 1994; Tabachnik & Fidell, 2007; Winer, 1971) provides the probability of exceedance for the null hypothesis that the means in all conditions are equal and does this by determining whether mean differences between conditions are large enough compared to variation of observations within conditions. Larger differences between condition means and smaller variation within conditions produce smaller p-values, hence reject the null hypothesis sooner. Figure 5.1 shows in the upper panel distributions of driving ability in five conditions running from, say, 20 mg, 15 mg, 10 mg, 5 mg, to 0 mg (left to right), that exhibit so much variation within each condition that the large

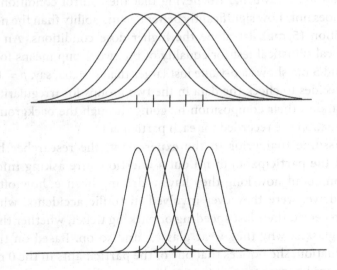

FIGURE 5.1 Five drugs dose conditions showing large within-group variance in driving ability (horizontal axis), obscuring group differences (upper panel) and five conditions showing smaller variance rendering group differences clearer (lower panel).

overlap makes it different telling to which condition John belongs if one did not know this. In the lower panel, the difference between the condition means is the same as in the upper panel but the conditions show less variation, thus producing less overlap. If one rejects the null hypothesis, this suggests that some or all the conditions have different means, so that we conclude that X influences Y. One may use *ad hoc* local statistical tests to compare pairs of conditions to find out which conditions differ.

Suppose the researcher finds in a plot she made that mean driving ability decreases with increasing drug dose but that the overall statistical test is non-significant. I assume she has no interests other than having her paper that reports the experiment and its outcome published. She knows that journals publish significant results and large effect sizes and tend to refrain from null-effect outcomes, like hers. She does not intend burying her research report in her desk drawer (Chapter 2). For the sake of argument, I assume she finds it hard believing that the control condition (0 mg) does not show significantly better driving ability than the next condition (5 mg), let alone the higher-dose conditions. An *ad hoc* local statistical test for equality of the two group means for 0 mg and 5 mg shows a *p*-value just larger than $\alpha = .05$, say, $p = .09$. She decides to check the data in the two groups for irregularities and studies their composition by going through the background information she recorded for each participant.

I assume that prior to the experiment, the researcher has asked the participants to fill out a questionnaire asking information about how long they have a driving license, how often they drive, were they ever involved in traffic accidents, when they received their last speeding or parking ticket, whether they wear glasses, why they wear glasses, and so on. Based on this information, she notices that one of the participants in the 0 mg group said he was far-sighted, and in the traffic-simulator made many judgment errors. She realizes he may have confounded the data if far-sightedness caused his low skill score. She decides to leave him out of the data. This raises the group-mean driving ability. In the 5 mg group, the participant scoring highest on driving ability reports driving almost daily to work and is longest in possession of a driving license. Because of possible over-qualification, she removes this person from her group, lowering the group mean. You may be reminded of the outlier example in Chapter 2.

The statistical test comparing the two adapted group means that are further apart after removal of the two cases produces $p = .059$, a result I made up for the example. A participant in the 5 mg group shows a funny pattern of judgment errors made during driving, with few errors in total, hence good driving ability, albeit all errors made in the final part of the driving test. The researcher suspects a concentration lapse and decides to remove this data line as well, lowering the group mean once more. Testing again for equality of group means produces $p = .03$, rejecting the null hypothesis of equal means in the 0 mg and 5 mg groups. She concludes that she had good reason to remove the three data lines and focus on the results comparing the control condition (0 mg) she expected showing better driving ability than the other groups and the 5 mg condition. She decides to offer some preliminary explanations in the Discussion section of the paper and submit it to a journal. She does not mention the first result, assuming she should have noticed earlier that she should have removed the suspicious data lines before the first statistical test.

I have two remarks. First, I suspect the example, even though artificial, describes in principle what happens multiple times every day. Second, I claim my imaginary researcher behaves sensibly. That is, anyone doing research using real data will recognize the urge one feels to dig further into the data when a result turns out differently than one wished. The desire to come up with positive or interesting results, sometimes fueled by mechanisms that boost competition and haste, may sometimes drive researchers to present their exploratory results as if they planned the statistical analyses rather than improvised them after having seen the initial results. The literature on QRPs does not mention often that the researcher may simply not master statistics well enough to fully realize what she is doing and what the consequences for the validity of her results are. The central thesis of this book is that insufficient mastery of statistics is a real bummer in research.

Two steps are reasonable when having produced results based on data exploration. One step is to use these preliminary results for amending one's theory and expectations derived from it as hypotheses and designing a new experiment to test the amended hypotheses. This might fit into a chain of events, systematically developing a theory, and using experiments to check a hypothesis. This strategy may be rather slow and painstaking, and not yield publishable results for some time, thus creating a situation that does not go well with a performance culture that demands quick results. The other step is to publish the negative confirmatory results along with the exploratory results and a discussion of what all of this could mean for one's theory and future research. This strategy requires an attitude of journals that tolerates not only success stories but also accounts of progress, no matter how painful yet realistic.

The experiment allows the researcher to manipulate conditions she hypothesizes to cause a particular effect, simultaneously controlling for all the conditions that also influence the effect but confound the influence of the target cause. For example, in the case about the causal effect of drug dose on driving ability, a confounding influence might be driving experience expressed in years, where more experience compensates deteriorating driving ability caused by a higher drug dose to an unknown degree. If drug dose and driving experience both influence driving ability, the researcher cannot disentangle their separate effects, unless she controls for experience when selecting the sample and thus isolates the effect of drug dose. Possibilities are to sample only participants of the same experience or take a stratified sample in which the distribution of different experience levels is the same in all conditions. Mellenbergh (2019) discusses various sampling strategies. When such sampling strategies are unavailable, statistical solutions can be used such as analysis of covariance (Pituch & Stevens, 2016).

Experiments are popular in research areas where it is feasible to manipulate conditions that are relevant to a particular phenomenon, and typically assign units at random to conditions to prevent systematic *a priori* differences between conditions that confound the effect of the manipulation. Experiments are impossible in areas where random assignment is unfeasible or unethical. Random assignment is unfeasible when studying, for example, the influence of a new arithmetic program on arithmetic ability, and the researcher cannot assign schools at random to a condition using this program and other schools to conditions using older programs when schools or another overseeing institution decide which program they use. However, the researcher may match schools volunteering to participate on general learning results, social economic status, and other possible confounders, and this way improve the *a priori* comparability of different conditions. However, this only compensates for non-random assignment when one first stratifies the population of schools on these confounders and then draws random samples in each condition, a possibility often unfeasible or unethical. Without matching, the researcher can record confounders and assess their influence using statistical methods such as analysis of covariance. This design typically represents a *quasi*-experiment (e.g., Shadish, Cook, & Campbell, 2002).

In the educational context and certainly in medical research, not providing a treatment—an arithmetic program for students weak in arithmetic or a life-prolonging medicine for patients suffering from cancer—that seems promising, to one group and not to others, raises ethical questions challenging the boundaries of what is acceptable in research. In general, one needs other research designs unlike (quasi) experiments at the expense of a loss of control over the validity of the results. I will use the survey methodology as an example, because it is so widely used in a great variety of research areas. Other examples where random assignment may be problematic include studying consumer habits, trends on the housing market, and effects of policy measures or laws on people's behavior. Prior to discussing survey research, I discuss policy to reduce the occurrence of exploratory research presented as confirmatory because this policy is mainly discussed in relation to experimental research.

PREREGISTRATION, REPLICATION, AND THEORY

The difference between exploration and confirmation is that between postdiction and prediction. Postdiction is like explaining yesterday's weather and prediction is using characteristics of the weather from the previous period to predict tomorrow's weather. The degree to which prediction with new data succeeds is a measure for the correctness of the theory one has for understanding weather systems. Data the researcher has already analyzed can generate hypotheses, but the researcher cannot use the same data to test the hypotheses these data generated; if you have already seen the outcome in the data, hypothesis testing does not make sense anymore. For hypothesis testing, the researcher must collect new data and test the hypothesis without further ado—that is, without looking at the data first to see if one should not rather adapt the data first or test another hypothesis. This would be comparable with predicting the weather; wait to see how the weather develops, and then adapt one's initial expectations if considered convenient. Yet, this happens all the time for many different reasons, and insufficient mastery of statistics is one of them.

Several authors (e.g., Lakens, 2019; Nosek, Ebersole, DeHaven, & Mellor, 2018; Rubin, 2020; Wagenmakers et al., 2012) discussed preregistration of hypotheses in a secured repository that time-stamps the entrance of the research plans including the hypotheses of interest, and all of this before observing the data. Examples of repositories are Open Science Framework (OSF) (https://osf.io), Dataverse (https://dataverse.org/), Figshare (https://figshare.com/), and Mendeley (https://www.mendeley.com/); see Nosek et al. (2018) for more possibilities. Preregistration is a case of placing bets until the race starts, but then all bets are off. However, after data are collected, some problems may appear that were unanticipated and sometimes the researcher can solve them without doing harm to the preregistration. For example, the researcher may find that some variables' distributions are skewed rather than normal, a circumstance she could not foresee, so that she needs to replace planned statistical methods with others accommodating the assumption violation. This is justified provided she re-preregisters this change of plans in the interest of transparency. Other adaptations may be unavoidable (Nosek et al., 2018), such as those caused by a smaller sample than expected, but move the research away from the ideal of preregistration. However, they may be necessary to continue research if the adaptations do not invalidate the results and additional preregistration maintains transparency.

Preregistration is often related to formal hypothesis testing (e.g., Wagenmakers et al., 2012), but is useful for other research strategies as well, if only to increase transparency about the researcher's intentions before the data are seen. In addition, whereas experiments are ideal for preregistration, any research design involving prediction may be preregistered, including survey designs that I discuss in the next section. Exploratory research also needs preregistration, if only to avoid researchers changing their mind during exploration and then continuing and reporting as if the research were confirmatory. One could argue that only confirmatory research must be preregistered, because by implication all other research is exploratory. However, I am afraid that for some studies this will lead to lack of clarity and endless debate one wishes to avoid, precisely because the reason to introduce preregistration was to limit the freedom researchers have moving from any sort of design to another design they claim is confirmatory when it is not. Haven and Van Grootel (2019) argued that qualitative research, using non-numerical data such as interviews, reports, photos, and eyewitness accounts for generating hypotheses and exemplary for exploratory research, should also be preregistered, just like other studies. The reasons are increasing transparency given that researchers' intentions are never completely absent at the start, informing colleagues about motivated digressions from the initial design, and disclosing one's theoretical stance with respect to the interpretation of the studied data.

In addition to reducing the likelihood of misinterpreting and misrepresenting exploratory research as confirmatory with claims about its robustness that are likely too strong, replicability is another motive for preregistering all research. Results obtained from hypothesizing after the results are known—HARKing for short (Kerr, 1998)—while wandering through a garden of forking paths (Gelman & Loken, 2014) without clear destination, stands less chance of successful replication than similar results obtained from straightforward testing of hypotheses on data that are new to the data analyst (Open Science Collaboration, 2015). Several authors (e.g., Fiedler & Prager, 2018; Stroebe, 2019; Trafimow, 2018) pointed out that in the human sciences exact replication is difficult, because experimental manipulations meaningful decades ago may have shifted in meaning due to social and cultural changes, thus losing validity. Other confounding influences may come from populations having changed on variables relevant to the manipulation and experimental procedures adapted to improve upon the initial experiment.

A statistically interesting issue is that focusing on replicating initially significant results introduces the regression effect (Galton, 1889), a selection effect that is active whenever you select cases with extreme variable scores, such as significance on the scale of the p value (Fiedler & Prager, 2018; Trafimow, 2018). If the null hypothesis is true, significant results indicated by $p < .05$ represent the 5% extreme cases at the low end of the scale. The regression effect predicts that extreme values on average regress to the mean of their distribution upon replication. Thus, one expects a sample of y initially significant studies to result in fewer significant replications, and the strength of the regression effect determines the magnitude of the decline. Examples of the regression effect, all other things equal, are students selected for re-exam based on failing the first opportunity (some will succeed even without additional preparation), and patients selected for therapy based on a positive intake interview (some will progress even without therapy).

The critics' point is that not finding the same significant result upon replication does not mean much. I suggest two alternative strategies. First, an initial study A and a replication study A_1 together constitute a sample of size 2 from the population of all imaginable replications of a study. The pair of outcomes (s = significant, n = non-significant) can be (s, s) or (s, n), and if one also replicates results that were non-significant the first time the study was done, the pair of outcomes is (n, s) or (n, n). A pair of outcomes is more informative than only one outcome, but if one would perform a larger series of replications, one would obtain a distribution of results for one study, facilitating a more solid assessment of the true effect. Second, rather than executing the same study of size N repeatedly, one could do the study using a large sample and estimate the effect size with greater precision lacking when sample sizes are small. A key question is whether researchers are prepared to perform many replications or one or a few large studies when journals publish only significant results and researchers put away non-significant results in their desk drawer where they pass into oblivion. Nuijten, Van Assen, Veldkamp, and Wicherts (2015) found that publication bias affects the influence replication has on the bias in the effect-size estimate. Effect-size bias may even increase, suggesting that publication bias may stand in the way of effective replication research including conceptual replication.

I expect that when a study is important enough to contribute to the development of a theory—a greater cause rather than an incidental study—researchers be prepared to replicate the study, whether its results were significant or not. Possibly, researchers are less prepared to do these things when a study concerns a topic that stands on its own but is not based on solid theory that is paradigm-driven (Nosek, Spies, & Motyl, 2012). That is, when the study's prospects are not resistant to stand the test of time, why bother doing a replication study? Who wants to spend time and resources replicating a study that may attract attention once, but probably has little future because it does not address a larger problem relevant to a better understanding of human behavior, the origins of a disease, or the existence of an elementary particle?

My discussion of the fishing expedition as either a well-prepared enterprise or a trial-and-error attempt not to return home empty-handed not only served to cheer prediction over postdiction but even more to argue that prediction must be supported by theory. To stand a chance of success, the scientific community must share a theory as promising, and this is different from an individual having an idea translated into yet another hypothesis to test. Doing research aimed at testing the theory invites replications or using larger samples, because it has a meaningful context many people share making the effort worthwhile. Many authors have discussed the necessity of developing theory for a science area to be fruitful; see, for example, Stroebe (2019), Eronen and Bringmann (2021), and Borsboom, Van der Maas, Dalege, Kievit, and Haig (2021). However, Ioannidis (2018) argued that most science areas are plagued by large numbers of significant first-study results that are false positives begging replication to clarify this, thus suggesting we are still remote from a theory-guided research practice.

Skepticism about the possibility of replications in the human sciences can improve attempts to found results on a stronger pedestal. Despite all problems in working with human beings as the objects of study, imperfect replication studies help to advance science albeit with less certainty perhaps than in the exact sciences. Not believing in replication research implies one also does not believe in the initial study; after all, one might reverse labeling study A_1 the initial study and study A the replication if one is prepared to regard them as pure replications of one another. Then, it does not matter which comes first. When A_1 is not a pure replication of A, the reverse is also true. What matters is that one tries to find evidence for an effect, meanwhile staying critical and prepared to reject a finding when evidence suggests this. Stroebe (2019) critically discusses the value of replication opposed to theory development in his field, social psychology.

A preregistration variety that fights the file drawer problem is Registered Reports (Chambers, 2013; Nosek & Lakens, 2014). The idea is that the journal editor assigns reviewers to assess the quality including the statistical power of the proposed research design before the data are collected. A positive review leads to acceptance for publication irrespective of the study's outcomes and discourages researchers holding negative outcomes behind that they expect the journal will not publish. The accepted research plan needs to be preregistered. Registered Reports also encourages replication research, which few researchers perform because of its small publication chances and therefore does not even make it to the file drawer. Registered Reports has the additional benefit of providing authors opportunities to improve their study based on review reports before they collect the data and protecting reviewers and editors from biased judgment based on knowing the results producing publication bias. Sijtsma, Emons, Steneck, and Bouter (2021) enumerate four benefits of preregistration: limiting researchers' degrees of freedom, ensuring replicability, revealing unpublished studies, and enabling registered reports.

Preregistration is still not common practice in several research areas (Baker, 2016). In a random sample of psychology articles published between 2014 and 2017, Hardwicke, Thibault, Kosie, Wallach, Kidwell et al. (2021) found that only 3% had preregistered. Bakker, Veldkamp, Van Assen, Crompvoets, Ong et al. (2020) studied the way predominantly psychology researchers preregister and concluded that many details about research plans were vague or missing and often deviated from the initial plans, sometimes indulging in HARKing after all. Gelman and Loken (2014) argued that they could never have formulated their most interesting hypotheses ahead of time, referring to analysis of existing datasets from politics, education, economics, and public opinion that other researchers already analyzed previously. They also make a plea for not straightjacketing statistical analysis by restrictions imposed by preregistration and note that replication studies are impossible in many fields where collecting new data is difficult or impossible, as with political election data. The authors recommend analysis of all data and comparisons when preregistered replication is impossible and rely on an increased awareness of researchers of the dangers of p-hacking and confusing exploration with conformation.

CONFIRMATORY AND EXPLORATORY SURVEYS

Because several authors advise limiting preregistration to hypothesis testing based on experimental designs, I give some attention to survey research and argue it should be preregistered as well. A survey is a study in which the researcher draws a sample from a population, asks a set of questions of the respondents, uses statistical methods to analyze the data, and generalizes results to the population of interest. Key to surveys is that the researcher cannot assign respondents at random to groups as in an experiment. The groups, such as based on education or religion, already exist. Many designs are available for sampling, and much effort concerns obtaining a sample that is representative of the population. For this purpose, one needs background variables or covariates on the candidate sample units. When the units are people, covariates may be gender, age, education level, income, religion, political party, and so on, depending on the goal of the survey. If the population consists of 27% Roman Catholics, 19% Protestants, 8% Muslims, 2% Jews, and 44% non-religious, and religion is an important confounder of the research question when not properly represented in the sample, the researcher draws a random sample from each category, such that the sample reflects the population distribution for religion. This is stratified sampling. Sampling is a statistical specialization of its own, and for some sampling schemes statistical methods need to be adapted (Mellenbergh, 2019).

The construction of questions and questionnaires is another specialized field for which a huge literature exists (e.g., Saris & Gallhofer, 2007). There are various ways to administer a questionnaire, for example, in writing, on a computer screen, in person, by telephone, via the Internet, and so on, and respondents may choose answers to questions from a set of pre-constructed answers or formulate the answer themselves (see the missing data example in Chapter 2). Respondents fill out the questionnaire or the interviewer does this based on the oral answers. Other data, such as biomedical data and data collected via smartphones or other wearables may be used in surveys. Moreover, surveys may compare different groups, such as age groups, in which case the research is cross-sectional, or follow the same group and measure the respondents repeatedly across time, in which case the research is longitudinal. Several other designs are possible. Statistical analysis is multivariate, and models may be complex, simultaneously analyzing the relationships between large numbers of variables. When the researcher did not previously identify groups as relevant to the research question, when available, covariates reflecting group membership may be part of the statistical model, which *ad hoc* incorporates the group structure in the statistical results.

Survey research lacks the strict control typical of experiments and usually produces large, multi-group, multivariate datasets that invite exploration rather than confirmation. Such datasets are heaven for QRPs, the variety of which runs in the dozens (Bouter et al., 2016; Fanelli, 2009; John et al., 2012; Steneck, 2006; Wicherts et al., 2016). The researcher's imagination limits the total number of QRPs possible. Notorious is leaving out observations from a dataset that seem to be unusual compared to most of the observations. Although *ad hoc* leaving out observations can be defensible, some researchers leave out observations only to make the remaining data look more consistent and produce results that the researcher believes seem more likely, desirable, or useful; see the outlier example in Chapter 2.

On the other hand, you need to realize that controlled experimentation involving the manipulation of independent variables is often impossible and limited in scope. Therefore, survey research is extremely important in the human sciences. The human sciences cannot do without it, but my point is that the richness of the data—the large number of variables, the often-complex subgrouping structure of the sample, and the availability of data collected at different measurement occasions—calls for discipline more than exploration. Wandering through a garden of forking paths (Gelman & Loken, 2014) trying numerous exits and finding out where they lead, may be irresistible and even wise to pursue, but then it is imperative to report the explorative nature of the data analysis, leading to the formulation of hypotheses to be tested in follow-up research.

Playing around with outliers, notable cases, and missing data (Chapter 2) refers to direct manipulation of the dataset qualifying as falsification, which obviously is different from the manipulation of an independent variable. Indirect manipulation occurs when one plays around with statistical modeling of the data to find the best model according to some criterion. This is different from the example I gave in Chapter 2 for multiple regression analysis using backward stepwise regression, which showed the unexpected large influence of sample size and algorithmic selection of predictor variables on the selected model. This was a one-step analysis resulting in odd outcomes. Here, I refer to trying many statistical models and selecting the model that gives attractive results. I use the multiple regression model again to show how this works.

Modern statistical models are complex in the sense that they contain large numbers of variables associated by means of mathematical functions. Let the model contain data for the variables Intelligence (resulting from an intelligence test) and Motivation (resulting from an application letter). Weights estimated from the data indicate their relative importance; see Chapter 2. I add the two products to represent another variable called Success in College. Thus, we have

$$Weight_1 \times Intelligence + Weight_2 \times Motivation = Success \ in \ College.$$

The model is very simple, and the two predictors explain Success in College imperfectly. The researcher can extend this model in many ways. One way is to include Gender explaining possible differences between models for male and female students. For example, Motivation may be more important for males. Then, the model becomes

$$Weight_1 \times Intelligence + Weight_2 \times Motivation$$
$$+ Weight_3 \times Gender = Success \ in \ College.$$

If you collected data on a large set of variables, many more than Intelligence, Motivation, Gender, and Success in College, and move different variables in and out of the model, you can construct many different models for explaining Success in College. This looks a little like moving observations in and out of the dataset, but the effects may be greater.

The example clarified that a model is a simplification of the truth, emphasizing salient features and ignoring details. The models I discussed contain a few variables relevant for Success in College but probably oversimplify understanding college success. Examples of other possibly relevant predictors are the personality traits of extraversion (the tendency to communicate with others) and conscientiousness (the tendency to act responsibly), the abilities of being able to concentrate for a long time and working spirit. In addition, situational factors play a role, such as whether one's parents were supportive and helped doing homework, whether one had brothers and sisters having academic ambitions, and the quality of the high school one attended. To find a reasonable model for college success, it helps to have a well-founded theory.

Models do not describe datasets perfectly well. In statistical parlance, the model does not fit the data, but if the misfit is large, the model is useless. In our example, gross misfit means that knowing a student's intelligence and motivation helps only a little to predict her chances of success in college: Students high on intelligence and motivation are more successful on average, but the mean differences with other groups according to the model are small. These small differences are due to the incompleteness of the model. To gauge the degree of misfit, several quantitative methods, such as the squared multiple correlation and the residual variance (Chapter 2), express the distance of data features as predicted by the model to the corresponding features in the real data. The squared multiple correlation usually is small and the residual variance large.

It is tempting to try different models, especially when the dataset contains many variables that could function as predictors in a multiple regression model or another model you could estimate and assess for its goodness-of-fit. Adding variables to the model or leaving variables out of it, trying different subsets of variables, adding products of variables to incorporate multiplicative or interaction effects, and adding background variables such as Gender and Family Background to account for possible differences between different subgroups may produce interesting results not anticipated. You may also consider using different kinds of models, such as non-linear models and models with latent variables. The array of possibilities is huge and so is the temptation to consider them. There is nothing wrong with exploring one's data if exploration is done in the open and reported in publications, so that the reader knows the model is tentative awaiting confirmation.

PREREGISTERING EXPLORATORY SURVEYS

Fishing expeditions are not necessarily a bad idea. One reason is that they may suggest amendments to an existing theory and new hypotheses to test. Focusing on subgroups in the sample and trying some alternative models are ways to stimulate one's imagination and creativity (Gelman & Loken, 2014). Two things matter when further exploring one's data. First, when publishing the results, the researcher must report the outcome of testing the original hypothesis. Next, she must state that what follows is the result of additional data analysis, exploring the data and clearly indicating the actions undertaken. Such actions may be the selection of subgroups, handling of outliers, notable observations, and missing data, and the probing of a couple of model variations. The result of the exploratory part of the publication is a suggestion on how to continue with the theory and the research testing the theory. Second, readers must not take my suggestion as a license for engaging in QRPs. Playing around with subgroups, data handling, and multiple models becomes a QRP when one presents the result of the umpteenth attempt as the outcome of testing a hypothesis that one formulated based on theory prior to designing the study and analyzing the data.

When the initial goal was confirmatory research, testing one or more hypotheses, and the research was preregistered, switching to exploration, generating ideas for future research is no problem. The publication of the research should make all this clear. Much research is explorative at the start, and I suggest it is preregistered as well. This way, a hidden switch based on an unexpected effect that is too good to be true not to publish is discouraged. The researcher should publish the unexpected and attractive effect as a possibly incidental finding based on exploration, but interesting enough to justify testing on a new dataset in a confirmatory context. Preregistration may help to focus researchers' attention on their initial intentions and discourage tacit digressions by making initial intentions public before the data are collected.

TAKE-HOME MESSAGES

Research areas lacking solid theoretical foundation are vulnerable to overinterpretation of non-zero results that appear in data without expectation or explanation.

Preregistration of research plans at least helps to increase awareness of the importance of confirmatory research.

Experimental research aimed at hypothesis testing is especially suited for preregistration, but other research designs including exploratory research must be preregistered as well to increase an open science culture.

NOTES

1 Similar to Table 2.2, make a table with 25 rows and 25 columns and insert the correlations in the upper triangle of the table. The number of cells in the table is $25 \times 25 = 625$. Delete the 25 cells corresponding to the same row and column numbers, representing the correlation of a problem with itself. In the remaining 600 cells, all correlations appear twice, so delete each duplicate. This leaves $625 - 25 - \left[\frac{1}{2}(600) \right] = 300$ unique correlations.

2 I used a two-sided t-test, $t = r\sqrt{\dfrac{N-2}{1-r^2}}$, with r the correlation, N the sample size, and $N - 2$ degrees of freedom.

3 This is only correct if the 300 tests are independent, but they are not because each variable appears in the 24 correlations it has with the other variables. Thus, the data of each variable are used 24 times, and this creates a certain degree of dependence among the various correlations. I have ignored this issue here and consider all correlations independent.

CHAPTER 6

Causes of Questionable Research Practices

THE LITERATURE ON RESEARCH integrity and ethics has devoted much attention to circumstances, such as publication pressure, competition for grant money, and desire to get tenure or promotion as causes to make haste, bend some rules, and sometimes cross an ethical border. In this chapter, I discuss the incomplete mastery of methodology and statistics as a cause of QRPs. I am referring to the mastery of the methodological rules and the statistical methods relevant to one's research area. It all seems so reasonable to assume that researchers master these skills and can apply them in their research without repeatedly making errors. However, the truth seems to be that too many researchers, without bad intentions, are unable to use methodology and statistics adequately. Do researchers accept experts' advice? What will one do when they mislead themselves, persevering in using inadequate methodology and statistical analysis, or deceive others when they give them incorrect advice based on faulty habit? Poor mastery of

DOI: 10.1201/9781003256847-6

methodology and statistics is a problem that calls for recognition and repair, not reproach. Why is poor mastery of methodology and statistics a problem in research practice?

My proposition is that researchers *need to use* methodology and statistics, but that methodology and statistics are not their profession; they practice skills in methodology and statistics only *on the side*. This causes QRPs. Publication pressure and other circumstantial problems increase the frequency of QRPs, but I assume QRPs have a base rate caused primarily by poor mastery of methodology and statistics. Most psychologists, cognitive scientists, sociologists, economists, biologists, biomedical researchers, and health researchers obviously specialize in substantive disciplines. This is what they studied in college and based on motivation and interest chose to do for a profession. How do methodology and statistic fit into an academic training?

Part of several academic training programs is the study of research methods that help to set up research designs typical of the discipline. Introductory classes in statistics and the use of the basic statistical methods practiced most frequently are obligatory in many disciplinary bachelor programs. The use of user-friendly computer software is a welcome addition to the theory of statistics, enabling the student and the candidate researcher to analyze a dataset with a beginner's knowledge of the programmed method, but without the need to understand how the method works mathematically. Whereas in the pre–personal computer era, say, prior to 1990, students were required to compute toy examples of the application "by hand"; when later given their own PC, they could let the machine do the computational work. And with the astonishing growth of computer memory and computation power and the quickly expanding library of statistical software, freely available and downloadable wherever an Internet connection could be established, students could let a computer do computations previously thought impossible.

The development of complex computer-intensive methods that require so many part computations that a human life would probably be too short to do everything by hand is another aspect of the computerization of statistics. As I write, such computations would still require minutes and sometimes hours using a powerful computer, but future developments promise even here an almost unbelievable reduction in computation time. To get a feeling of what I mean, I recommend looking up a few papers published in one's research area from the times before 1960 and let the small scale at which experiments and survey research was done surprise you. Then, realize that people had to do all the computations by hand, and that research assistants, sometimes called *computers*, meticulously checked the results by doing them all over again. Until the 1970s, only big stand-alone computers the size of small trucks were in use, and in many areas, research looked different from what it is today. Researchers used punch cards for software instructions, one 80-character line for each card, hand them in at a counter and come back the next morning to pick up a stack of paper showing the output of the statistical program. An error caused the program to break down and print a "fatal error" message, and the cycle started all over again. Anyone interested in the amazing conquest of science by the computer should consult Gleick (2011) and Dyson (2012).

With all the blessings computerization has brought us comes the diminishing need to understand the intricacies of the statistical methods one uses to do the computations. Add to this decrease in understanding the growing complexity of statistical methods, primarily fueled by what computers render computationally feasible. In this chapter, I will argue, helped by some of the greatest minds psychology has produced, that one should not believe all too easily that researchers understood small-scale statistics before 1960 very well. My point is that now that machines have taken over, profound understanding is further away than ever. This does not mean that researchers trained and motivated in a substantive area cannot be excellent statisticians as well; they can, and several are! However, in addition to these statistically well-versed researchers, there are quite a few that practice methodology and statistics on the side next to their main profession. And who can blame them? After all, they want to know what makes juveniles engage in illegal activities, why consumers are attracted to a product, and how a virus spreads among a population, but this does not imply that they are also interested in the derivation of the F-statistic used in analysis of variance or the step-by-step estimation algorithms perform in multilevel analysis.

The knowledge substantive research areas requires is very different from the knowledge needed to understand statistical reasoning. In addition, for many people statistics is more difficult to understand than other topics. Just think of the difficulty young children experience when learning and practicing arithmetic in primary school, students when trying to master algebra and geometry in secondary school, and college students when following statistics classes in academic training. When they become researchers, they usually did not follow more statistics classes than their fellow students who left university for jobs outside academia. Where then do they pick up the statistics skills needed to do research? Many possibilities exist: By following additional and more advanced statistics courses, by listening to their supervisors and colleagues when they are PhD students or junior faculty, by attending crash courses given by experts in complex methods such as multilevel modeling, structural equation modeling, and meta-analysis, and by reading a textbook on the subject. I applaud all these activities, but I will argue that for many researchers, these possibilities for acquiring more statistical skills are not enough to engage safely in statistical analysis of the complex datasets collected nowadays.

One of the typical characteristics of statistical reasoning is that it follows a strict logic. Another characteristic is that is deals with probabilities requiring thinking in terms of uncertainty. The final characteristic is that computational results are often unpredictable if not counter-intuitive. Together, logic, uncertainty, and counter-intuition provide a dish that is hard to digest. In what follows, I will provide a few examples of the logic, the uncertainty, and the counter-intuition so typical of working with probabilities, distributions, and multivariate data. You will find that the answers to the problems are not only unexpected but downright unintelligible. After the examples, I will lean heavily on the work of Amos Tversky and Daniel Kahneman to explain why statistics is difficult, also for statisticians, and why researchers not completely certain of their cause do best to consult a statistician when analyzing their data. Finally, I will explain that consulting experts is great way to avoid QRPs.

STATISTICS IS DIFFICULT

I discuss two toy probability problems that have counter-intuitive solutions. Following each problem, I discuss realistic statistical problems that share some formal features and represent some issues researchers encounter in their data practice. The examples serve to clarify the misleading nature of statistics.

The Birthday Problem

The first example is the birthday problem, well known from introductory probability courses. Select 23 persons at random from the population; what is the probability that at least two have their birthday on the same date, except 29 February? Now, it is essential that you do not read on immediately, but first try to give the answer. You will probably find that the problem is hard to tackle, and resort to guessing based on your understanding of the problem. I predict that many readers will say that the required probability is small, rather close to zero. After all, a year has 365 days, and 23 persons are much fewer than 365 days. The truth is that the probability is just over .5. This is quite a large probability, comparable with obtaining heads (or tails) when tossing a fair coin. If the number of people in the room were 50, the probability equals .97. This means that with this small number of people relative to 365 days, it is almost unavoidable to find at least two persons who have their birthday on the same date. A caveat concerning the explanation of the problem is that it does not provide a revelation to many people. This is typical of the point I am making: Once you know the answer, even if you know how to obtain it, the answer still looks mysterious. The formal solution is the following.

I arrange the birthdays of n persons in an array consisting of 23 entries, each entry being a date. For each different random sample, I get something like:

| 23 Jan | 28 Oct | 2 May | 8 May | 12 Dec | ... | 8 Apr |

It is important to notice that dates can appear more often in one array, and we are after the probability that this happens. Assuming all dates are equally likely, and 29 February is excluded, there are $T = 365^n$ different arrays like our example. Each outcome has the same probability. I find it convenient to turn the birthday problem "around" and ask for the probability that *no two (or more) persons* have their birthdays on the same date or, equally, that n persons have their birthdays on n different dates. For the first person, there are 365 different dates available, for the second person 364 dates except the date the first person gave, and so on. Hence, the number of different outcomes containing n unique dates equals $U = (365)(364)(363)...(365 - n + 1)$, and the probability P that no two persons have their birthday on the same day equals U/T. For $n = 23$, one finds $P < .5$, so turning to the original problem, the probability that at least two persons in the group have their birthdays on the same day exceeds .5. Computations for other values of n follow the same line of reasoning.

The example makes clear that events may have a much greater probability than one expects, and my experience is that people find this difficult to believe when their intuition initially told them otherwise. Even when confronted with the computations, initial belief is stronger than reason, it seems. In statistics, probability plays the main part and often is different from expectation. In the example, key to beating intuition is the realization that one must not compare 23 people to 365 days—that is, compare the numbers 23 and 365—but rather the number of unique pairs of people.

With J persons, there are $\frac{1}{2}J(J-1)$ unique pairs. You may check this by writing down the unique pairs for $J = 3, 4$ and so on. For 23 persons, there are 253 unique pairs. The actual computation of the probability in the birthday problem is more involved than, say, dividing 253 over 365, and difficult to find out for yourself.

You may object to the birthday problem that it is a contrived example that mainly serves as a puzzle demonstrating how little people understand of probability but without practical consequences. However, I contend that such problems demonstrate convincingly that logical reasoning problems are the hardest we face and that we should admit this without restraint. Almost everybody struggles with these problems. Moreover, the problem's logic generalizes right away to everyday statistics, as the next example extended from Chapter 4 shows.

Testing Multiple Hypotheses

I replace 23 persons with 23 variables and ask how many you expect to correlate significantly when all the true correlations are zero. We considered this problem already in Chapter 4 for 25 0/1 scored arithmetic problems. Twenty-three variables share 253 correlations, and with significance level $\alpha = .05$ and assuming independence, I expect $253 \times .05 = 12.65$ correlations significantly different from 0 in the sample. Across samples, I expect a distribution with mean 12.65 (also, see Table 5.1, for 25 variables). To bend the example toward the Birthday problem a bit more, I ask for the probability that one finds at least one significant result, two, three, et cetera. If I define a significant result as a success, I am dealing with 253 trials where each trial has a success probability $\alpha = .05$. The probability of $X = x$ significant results, given $n = 253$ independent trials with success probability .05, follows a binomial distribution, which I approximate by a normal distribution; see Appendix 6.1. Table 6.1 shows results for $x = 1, ..., 24$.[1]

From Table 6.1, I conclude that finding between 7 and 18 significant correlations should come as no surprise, even when all

TABLE 6.1   Probabilities (Prob) of Finding At Least x Significant Correlations at $\alpha = .05$

x	Prob	x	Prob	x	Prob	x	Prob
1	.9998	7	.9620	13	.5173	19	.0458
2	.9994	8	.9313	14	.4032	20	.0241
3	.9983	9	.8844	15	.2968	21	.0118
4	.9958	10	.8182	16	.2055	22	.0053
5	.9906	11	.7324	17	.1334	23	.0022
6	.9804	12	.6300	18	.0809	24	.0009

true correlations are 0 and there is absolutely nothing worth finding, unless you expect no relationships and find that result worthwhile. Even when all true correlations are 0, sample results give us plenty of opportunity to speculate about a relationship we did not expect but that looks interesting now that it presents itself. Given that all true correlations equal to zero is unlikely, I expect the number of significant results in real research is larger than 13.

Climbing to the Moon

Here is another apparently contrived problem with practical consequences for the data analyst. Fold Saturday's edition of your national newspaper (say, 1 cm thick), then fold it again and keep folding. When is the pack of paper thick enough to bridge the distance from here to the moon? Obviously, you cannot do this exercise in practice because folding very quickly becomes physically impossible, but what counts is the intellectual exercise. Again, I ask the reader to try giving the answer first, and like the previous exercise, I expect many of you to guess using your best understanding of the problem. This time I predict that many readers will say that the number of times one must fold the newspaper is enormous, given the thickness of the newspaper and the distance to the moon, which due to its elliptic orbit varies between two extremes with a mean equal to 384,450 km or 38,445,000,000 cm. The truth is that the number of times one must fold the newspaper is very small: Folding 35 times is too little, but 36 times is too much, and again I expect knowing this does not help you to understand the problem ("Now that you mention it, I see that ..."). The formal solution to the problem is the following.

You may think of the number of score patterns 25 0/1-scored arithmetic items enable (Chapter 4), which proved to be enormous, $2^{25} = 33,554,432$, to be precise. It helped to start with one problem with two outcomes; add a second problem producing four score patterns of two 0s and 1s in total, add a third problem producing eight three-digit score patterns, and so on. The point was that with the addition of each next problem, the number of possible score patterns doubled. This mechanism is precisely what explains the folding of the newspaper: Each time you fold the newspaper, it becomes twice as thick as it was. In the first step, 1 cm becomes 2 cm. In the second step, 2 cm becomes 4 cm. Then, 8 cm in the third step, 16 cm in the fourth step, and so on. After x steps we have thickness equal to 2^x cm. Now the question is which x produces outcome 38,445,000,000, the distance to the moon in centimeters. The problem therefore can be written as: Solve x from $2^x = 38,445,000,000$. There is no exact integer solution for the number of times one folds, and one finds that 35 times is far too little, and 36 times is far too much. Just try 2^{35} and 2^{36}, and you will see.

Exponential Growth in Multivariate Data

If you study one variable, you can consider its distribution and the distribution's properties, such as the mean, the variance, the skewness (whether the distribution leans to one side more than the other side), and its flatness or peakedness (also known as kurtosis). Because many variables are discrete, and because most samples are relatively small, distributions derived from the data are histograms. A bar's height in histograms represents the frequency of the value it represents. An example of a discrete variable is people's age expressed in years or other units. The choice of the unit depends on the researcher's intentions and the requirements the research questions pose. For many research questions, it is enough to express age in years. Many statistical methods treat variables as if they are continuous, even if they are discrete. For example, the much-used method known as factor analysis (e.g., Pituch & Stevens, 2016) assumes that variables are continuous but is usually applied to discrete data. A common example is rating-scale questions from questionnaires, often having only five ordered answer categories scored 1, 2, 3, 4, 5 to indicate a higher level of agreement with a statement (Dolan, 1994).

What happens if one considers two or more variables simultaneously? The purpose is to study the association between variables from the simultaneous distribution of these variables. For two variables, say, happiness and income, the simultaneous distribution looks like a hill (Figure 6.1), with the values of the two variables on orthogonal dimensions (like length and width), and frequency of combinations of values represented as height. Association is visible from tendencies in the hill looked at from

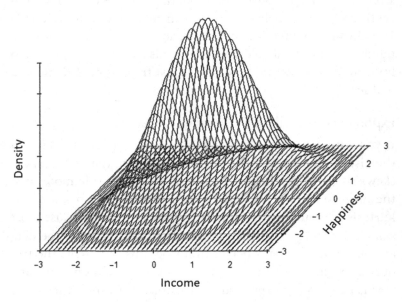

FIGURE 6.1 Example of a bivariate distribution for Happiness and Income.

above. For example, the crest of the hill in the figure suggests a positive association between income and happiness; happiness tends to increase with income. This does not mean that increasing income produces happier people, let alone that happiness produces higher income. It only means that both variables tend to increase simultaneously. We may ask more-involved questions, for example, whether the relation between income and happiness varies with degree of urbanization and religion. Simply *looking at* simultaneous or multivariate distributions becomes problematic when more than two variables are involved, but algebraically this can be done smoothly using multivariate statistical methods. I will discuss a small example with four variables to show some of the difficulties involved.

In addition to income (say, six ordered categories) and happiness (I assume five ordered categories), I consider degree of urbanization (four ordered categories, based on number of people per square mile), and religion (five unordered categories). Crossing these four variables yields a four-dimensional cross table with $6 \times 5 \times 4 \times 5 = 600$ cells, each cell representing a unique combination of four scores. Each person has four scores and is in one cell only, so it would take 600 persons to have one observation in each cell. Obviously, some cells are more prevalent than other cells. For example, in the Netherlands, different religious groups have highly varying prevalence in the population. Clearly, for realistic overall sample sizes, many of the cells remain empty and many others contain one or a few observations. Empty and low-frequency cells provide a problem for many important statistical methods. Without clever statistical solutions, computations are impossible, estimates of certain quantities lack precision, and several groups are absent, so that results are flawed.

Modern statistics puts much effort in developing complex models that "smooth" the data based on assumptions about the sampling model and the variables' distributions. These methods serve to compute the models without making serious errors but cannot entirely make up for a largely empty cross table.

The example shows how quickly cross tables that are at the heart of multivariate statistical methods become too big to handle well. The number of cells grows exponentially with each new variable. This also means that the table misrepresents many groups unless the sample size is enormous. Even with stratification, the sample must be huge to have enough observations for each group. The example demonstrates the exponential growth of cells in cross tables as the numbers of variables and the numbers of categories per variable increase, and exponential growth goes at an enormous pace.

INTUITION RULES, RATIONAL REASONING LAGS BEHIND

When confronted with QRPs throughout my career, I often wondered what their cause was. I could never figure this out beyond speculation. When among one another, statisticians talked a lot about the statistical problems they encountered in their daily consultancy. In addition to the errors researchers make that I encountered myself, these discussions told me a lot about colleagues' experiences, but very little about the reasons why researchers make so many mistakes and seem to misunderstand statistics so thoroughly. It seemed that researchers underestimated the difficulty of statistical methods and statistical reasoning. I had devoted my career to statistics and still found it difficult, so how could experts in different, non-statistical research areas not be more careful? A colleague from another discipline once asked me to read a concept chapter of one of his PhD students and give my comments. The topic matched my skills well, and when I read the chapter, the huge number of errors and even blunders I encountered nearly swept me off my feet. Trying to bring the message cautiously and without offense, I limited my comments to some general issues and offered a helping hand. My colleague never returned to his request and my response, and I decided to let it rest. Perhaps I must blame myself, but when the dissertation appeared, I noticed that almost all the serious problems I identified were still there.

Why did my colleague turn down my offer to help? Perhaps he was only looking for a pat on the shoulder and endorsement, but not my comments, let alone my involvement. Perhaps I chose the wrong words; I do not know. I remember thinking: Why does he take these risks letting his PhD student muddle through, both not being experts in statistics, when even I, at least an experienced applied statistician, find what they are doing difficult and offer help? Experiences like this accumulated for many years, but then the Stapel affair became a catalyst in my thinking about clumsy statistics. Having started as Dean in September of the previous year, in the summer of 2012, when I was on vacation and contemplating the affair that I had tried to manage now for some ten months, one night I started writing down my experiences with failed consultation and the kind of statistical problems involved and tried to understand what happened. I already had read some literature on QRPs. In the next months, I rewrote what I had concocted many times and started presenting the result to audiences who were eager to know more. I asked many colleagues coming from varying backgrounds to read different drafts of the manuscript and provide me with their comments, but all the time I felt my story was incomplete and did not rise above the anecdotal level. The problem was that I did not have a believable explanation for the kind of risk-taking behavior researchers engage in and that I had seen so often, and that this was necessary to make the manuscript worthwhile reading. In the spring of 2013, with the worst extravagancies of the Stapel affair behind me, I picked a book in an airport's bookstore. I started reading on the plane and soon could not stop reading. Already the first few pages of Daniel Kahneman's *Thinking, Fast and Slow* provided precisely what I had been looking for in previous months.

For a quick understanding of some of the ideas Tversky and Kahneman (1971, 1973, 1974; Kahneman & Tversky, 1982) developed and tested and that are relevant to statistical reasoning, let us have a second look at the birthday problem. A general principle of the workings of our mind is that it entertains a strong tendency to understand the world, even if understanding draws from incomplete information, either correct or incorrect. The main purpose is to construct a coherent story of a particular phenomenon, not necessarily a correct story. A special feature of these processes is that there is no active search for additional information; we use the available information without asking for more. All of this happens just like that; given a particular stimulus, such as a question, our mind associates memories that are activated without creating awareness of the process, and quickly comes up with a result that may be reasonable given the truth, or nonsensical, or anything in between. What matters most is that we feel good about the result. Briefly, this is what our System 1 does. System 1 is a metaphor for a rather diffuse set of processes and operations that run without us noticing. A way to think about the workings of System 1 is that they represent intuition. System 1 is fast and inaccurate. It is imperative to understand that each of us without exception uses System 1 as a first response to a stimulus that is intuitive. We make many mistakes by doing this, but for most occasions in everyday life that are not crucial, responding approximately adequately is enough to get you through the day without big accidents. We jump to conclusions all day and muddle through life.

Because initially, we always respond intuitively, you also greet a statistical or another formal problem by activating System 1, that is, your intuition. To understand what makes the birthday problem so difficult, this is what typically happens when confronted with it. You read a problem. Only if it is very simple will you give a direct answer to it; otherwise you will activate your System 1. The birthday problem is very difficult. This means you will follow your intuition and jump to conclusions. The target is a coherent story, not a correct story. Interestingly, intuition works two ways:

- Based on *heuristics*; this kind of intuition makes you respond using cognitive automatisms that replace the difficult but correct question that is impossible to answer with an easier but incorrect question that is answerable. This is the process of substitution. It is satisfying to System 1 because it produces an answer but puts you on the wrong track. System 1 does not care.

- Based on *experience*; this makes you do approximately the right things before you have even started thinking about a well-founded answer. Experience works as a kind of warning automatism: You do not know how to solve the problem, but you have a hunch that it might be difficult. This can have the effect of postponing a response—you withdraw and think the problem through, however, without guaranteed success.

Thinking the problem through is not part of System 1 but of System 2. Whereas System 1 is your autopilot, System 2 means hard work and a lot of effort. System 2 can deal with complex problems and analyze them rationally, but this does not come naturally. Kahneman (2011, p. 20) gives the example of the arithmetic problem "17 × 24" representing the maximum of mental effort we can deal with from the top of our heads. System 1 tells you that this is a multiplication problem, that you can solve it, and that 123 and 12,609 cannot be the correct answer, but when presented 368 you could not tell with certainty whether this is the correct answer. You can force yourself to solve the problem by heart, but this costs a lot of effort when you must remember how to multiply, keeping a lot of material in short-term memory, keeping track of intermediate results, and running several ordered steps to arrive at the result. You must do all of this deliberately while investing effort, and in an orderly fashion. Even then, a correct result is not guaranteed; you may make computational or memory errors, and even use an incorrect solution strategy that leads you astray. When the problem is too difficult to do by heart, you may use a notepad or a calculator but even then, you may make errors producing an incorrect answer. The way System 2 operates suggests it is in control and corrects the incorrect intuitions System 1 makes. However, this is not what happens. System 2 is more like a facility that you can sometimes choose to operate, but if the solution System 1 suggested is convincing enough, not necessarily correct, System 2 may not become active at all. Let us go back to the birthday problem to see how this works.

The birthday problem was:

> Select 23 persons at random; what is the probability that at least two have their birthday on the same date (except 29 February)?

Because this is an extremely difficult question, it should come as no surprise that at first sight nobody knows the answer immediately. This remains true if you would grant yourself the opportunity to sit down and try to solve it rationally—that is, using System 2. What almost everybody does instead, without knowing it, is give in to System 1, which operates a *heuristic* process. Concretely, System 1 replaces the difficult question with a simpler question that you know how to answer. For example, you might answer the question:

> Do I know people who have their birthday on the same day?

An answer that is likely to come up immediately is: No (or ... just one pair ...); hence, the requested probability is small. Perhaps you have a birthday calendar at home or at work and an immediate mental image of some of its pages you retrieve from memory suggest that most names appear scattered but rarely in clusters on the same dates. Again, the answer that pops up is no. You may recognize the powerful heuristic of substitution. For most questions or problems that surpass "17 × 24" in complexity, you do not have ready-made and correct answers. Substitution provides you with an answer to a question or a problem that is too difficult for your cognitive machinery. Therefore, intuition by heuristics takes over and provides you with an answer to a question or problem similar but different from the original question that makes you feel good and confident, not inclined to activate System 2, and indulge in a long, effortful, and uncertain search for the correct answer.

How does a statistician distinguish herself from using intuition by heuristics, or does she? Expect no heroine's story. When confronted with the birthday problem, a statistician, an expert, without thinking consciously, recognizes a cue in the problem, which provides her with access to information stored in memory, and this information produces an answer. In the case of the birthday problem, the cue may be the simple realization that this kind of problem is never self-evident, has several layers, and is deceptive. Another cue is that you must consider pairs of people, triplets, quadruples, and so on, and that this is complex and the outcome unpredictable. These possibilities show that the answer need not be final and correct. To solve the problem, the statistician must move to System 2. What matters here is that the expert recognizes features that make her cautious without being able to solve the problem right away.

The groundbreaking insight here is that *intuition is recognition*, not some kind of magic as some choose to believe. But why is *recognition* available to an expert and not to other people? The question is not entirely correct. It should be: Do non-experts also recognize cues in problems? The answer is yes, but different problems and different cues require different levels of training. When a dog once bit you, you will respond quickly and emotionally to the sight of an unknown dog running toward you, even without having given it the shortest moment of thought. The sight of a dog running toward you is enough to retrieve the cue from memory that tells you to run away or something similar (but perhaps very ineffective). This simple situation gives rise to a simple cue that requires no mental effort to retrieve from memory, but the birthday problem is very complex, and recognizable cues are the result of years of training in studying and using statistics.

The statistician, who is an expert—that is, well-trained—in manipulating probabilities, distributions, sampling, and especially in counterintuitive results originating from enormous numbers of different patterns typical of even small numbers of variables, has the ability of intuition by *experience*. Typically, this works as a warning signal to back off, take more time, not jump to conclusions; that is, ask the researcher seeking consultation for more time to consider the problem thoroughly. Everyone asking a statistician for help with a nasty statistical or data-analysis problem must have observed that the expert will ask you many questions and at the first meeting provide few if any solutions. The large array of questions serves to help the expert understand the problem and the question, and ascertain whether an answer to the question solves the problem or whether a different question is adequate when you asked a question that was not entirely adequate. Not providing an immediate answer shows that the problem is difficult and takes time to understand, and asking for a timeout means that the statistician needs time to set System 2 in motion. The statistician cannot do this just like that, and must take deliberate action, invest a lot of effort, and work in an orderly way that may need design because it is not standard. And all of this may fail when the effort is too large, the solution is wrong, or when it simply "feels good," thereby satisfying System 1. System 2 is no panacea and besides being slow, it is lazy.

Does this mean that researchers who are not trained statisticians are not bright enough to use statistics to analyze their data? The answer is no. They are simply inexperienced when it comes to understanding and using statistics. Even though many researchers received training in basic statistics and data analysis during their bachelor's and masters' programs, the knowledge acquired is only the beginning of a proper statistics training and will not make them experts in statistics. You have a basic idea of the statistical tools that are in the statistical toolbox, but that does not prevent you from making errors when choosing statistical methods and using them to analyze your data. For that, you need experience in statistical thinking. Like all formalized and logical thinking, statistical thinking is difficult and often steered by processes that are complex, producing counter-intuitive results. When you are unaware that you are using statistics *on the side* without having acquired the experience needed to realize at least that using statistics is like walking on thin ice, accidents will happen. These accidents can take the shape of QRPs. One of the characteristics of System 1 is the overconfidence people have in the solutions they embraced unconsciously and the resistance of these solutions to amending. For this, you need System 2.

If anything, QRPs happen when System 1 is in control and intuition by heuristics determines the solution to problems. System 1 also rules statisticians' minds, but experience grants them the possibility to find the brake and let intuition by experience buy time before accepting a solution. After that, System 2 is activated, but without guarantee of success. Whereas System 1 does not involve deliberate action, System 2 does. It is interesting to realize therefore that System 2 is also the habitat of FFP. It is hard to imagine that one could fabricate or falsify data or plagiarize a text other than by intent. Not having realized what one was doing when FFP is involved is unbelievable. System 1 simply runs, and intention is not involved, but System 2 involves intent.

I noticed that the distinction between System 1 and System 2, or fast and slow thinking, easily produces confusion and even resistance among readers thinking that I imply that System 1 always fails solving problems whereas System 2 always succeeds. Here is a synopsis of what I meant to convey. System 1 helps us respond to *all the situations* we encounter both in daily and professional life. First, there are the situations the response to which appeals to routine not posing problems, such as shopping in the supermarket and driving a bike or a car. Second, there are the situations where we do not have a solution available because we lack knowledge and experience. System 1 comes up with solutions that are coherent and not necessarily correct but hopefully approximately correct and practically useful. However, someone possessing knowledge and experience relevant to the more-complex situations may respond differently because *recognition* of relevant problem characteristics activates System 2, which allows using the knowledge to respond correctly. Unfortunately, System 2 is no guarantee for success because it is slow, lazy, requires much energy, and tends to fall back on the fast and energetic System 1. When faced with more-complex statistics problems, researchers who lack knowledge and experience are drawn into System 1, leading them astray. It is here that they need a little help.

In previous chapters, I mentioned the extreme competitiveness of modern-day science as an environmental strain on researchers driving several of them to QRPs. They do this not because they want to but because so much depends on publishing many articles in top journals, acquiring large amounts of grant money, and supervising large numbers of PhD students (Miedema, 2022). For young people, tenure often depends on such accomplishments and for senior researchers, more articles, more money, more coworkers, more sponsored invited lectures, better access to interesting people and organizations that distribute grants, more fame, admiration from colleagues, and prizes emphasizing excellence all depend on high scientific productivity. One may argue that scientific accomplishment typically produces this kind of return on investment, and I concur. Good results call for recognition. For this book, the question is what tough competition does to the use of statistics when non-experts cannot get around it in their data analysis, while at the same time, statistics is difficult, and results are counter-intuitive, constantly misleading researchers' System 1. When you are under pressure to do something at which you are not an expert, QRPs are unavoidable.

Over the past decade, I have given several lectures that follow the structure of this book to academic audiences in several countries. I usually start with some personal experiences with FFP, and the Stapel case guarantees catching people's attention. Of course, this old trick serves to make way for some messages that I hope audiences will recognize and will win their approval. What I noticed is that people much appreciate the personal experiences and listen with interest to the part on how statistical thinking misleads almost all of us and together with unreasonable performance pressure may be regarded as the cause of QRPs. However, I also noticed each time that the discussion afterward is rarely about researchers' failure to use statistics correctly but rather about everything else with respect to FFP and QRPs, and the way we can fight them.

Kahneman (2011, pp. 215–216) mentions a case where he correlated the performance rankings of 25 financial advisers collected in seven years for each pair of years and found a mean correlation of .01. He took this as evidence that there was no persistence in skill, but that luck rather explained the rankings. When presenting the results to the executives and the advisers, their reaction was not one of disbelief, but the message rather did not seem to come through. Kahneman concludes that the executives and the advisers simply held very firm beliefs about their skills and that a finding that contradicted their belief could not shake that belief. I speculate that for researchers the same mechanisms are in place and that someone explaining that statistics is simply too difficult to practice *on the side* without a high risk of QRPs will be listened to but not believed. After all, the message concerns all of us and is painful, and System 1 will immediately defend its territory by suggesting likely alternative explanations. Interestingly, researchers readily accept extreme performance pressure as cause of QRPs, perhaps because this cause is situational, located outside the individual. Much as I agree with this point, thus taking external causes of QRPs seriously, the workings of our mind is another cause of QRPs and the cooperation of external and internal causes is a toxic mixture we have to recognize and resolve.

Fortunately, I have met quite a few researchers who are able statisticians. All of them have a deep interest in statistics and often in other topics not directly related to their main specialization, and I expect they combine talent with hard work and training to become experts in more than one area. I have also met quite a few researchers who recognize their own limitations, accept that you cannot be an expert in everything, and seek advice and cooperation with statisticians they happen to know. However, the fact that statistics is difficult and constantly produces counterintuitive results, that researchers have not acquired the experience to respond at least cautiously, and even worse, that our mind is not at all susceptible to the truth but rather seeks to produce a coherent story that makes us feel good and suppresses attention, concentration, and effort, does not help. Many researchers are not experts in statistics, and frequently fail to use statistics adequately without ever knowing this. A consolation may be that statisticians like me continue to find statistics difficult as well.

TAKE-HOME MESSAGES

Many researchers practice statistics on the side, not acknowledging they must deal with logical rigor, uncertainty, and counterintuition. This ignorance underestimates the need for experience and causes QRPs.

External sources driving competitiveness among researchers by requiring large numbers of articles, grant money, and PhD students increase the occurrence of QRPs already caused by lack of statistical skills.

Query: Imagine a world in which researchers master statistics and data analysis well. Will QRPs be a problem?

APPENDIX 6.1: BINOMIAL PROBABILITIES AND NORMAL APPROXIMATION

The probability of x successes (here, significant correlations) given $n = 253$ independent trials (here, 253 different statistical tests on 253 different correlations) each with probability of $\alpha = .05$ equals

$$P\left(X = x \mid n, \alpha\right) = \binom{n}{x} \alpha^x \left(1 - \alpha\right)^{n-x} = \binom{253}{x}.05^x \left(1 - .05\right)^{253-x},$$

in which $\binom{n}{x}$ is the binomial quotient that counts the number of different patterns of n trial outcomes of which x are successes, and

$1 - \alpha$ is the probability of failure (i.e., a non-significant correlation). I compute the probability of observing at least x successes, which is equal to the probability of observing no more than $n - x$ failures. For the first probability, I need to compute probabilities for $x, x + 1, x + 2, \ldots, n$, and starting with $x = 1$, this means computations for $1, 2, 3, \ldots, 253$; that is, 253 probabilities and adding them,

$$P\left(X \geq 1 \,\middle|\, n = 253, \alpha = .05\right) = \sum_{x=1}^{253} \binom{253}{x}\left(.05^x\right)\left(.95^{253-x}\right)$$

This is the probability of finding at least one significant correlation. Alternatively, I compute the probability of observing 0 successes, and subtract the result from 1; that is,

$$P\left(X \geq 1 \,\middle|\, n = 253, \alpha = .05\right) = 1 - P\left(X = 0 \,\middle|\, n = 253, \alpha = .05\right) = 1 - .95^{253}.$$

The latter result certainly is less involved than the first and produces the same result, but both suffer from computational inconveniences caused by large n and small α (although the final result is readily seen to be very close to 1, using the calculator on your laptop or smartphone. Feed .95 to it, press the x^y button, and then feed 253 to your calculator; this produces $.95^{253} \approx .000002312$, which is nearly 0. Hence, $P(X \geq 1 \mid n = 253, \pi = .05) \approx 1$, showing that you can hardly avoid finding at least one significant correlation).

The binomial distribution used here can be approximated by means of the normal distribution with mean $\mu = n\alpha = 12.65$ and variance $\sigma^2 = n\alpha(1 - \alpha) = 12.0175$ and standard deviation $\sigma = \sqrt{12.0175} \approx 3.4666$. Without going into details and accepting that this normal distribution well approximates the binomial, I determine the area under the normal curve that corresponds with $P(X > x - .5 \mid \mu = 12.65, \sigma = 3.4666)$. Subtracting .5 from x is a continuity correction. For $x = 1$, I therefore use standard score

$$z = \frac{(x - .5) - \mu}{\sigma} = \frac{.5 - 12.65}{3.4666} \approx -3.5049.$$

A table for the standard normal distribution will show that $P(Z > -3.5049) \approx .9998$, which comes close to 1. The difference with the exact computations is due to the continuous normal approximation of the discrete binomial distribution, but I will let this nuance rest.

NOTE

1 I used statistical package R to compute the area under the normal curve.

CHAPTER 7

Reducing Questionable Research Practices

I F YOU ASK A statistician how to reduce the occurrence of QRPs, chances are that she comes up with a statistical solution. This is not to suggest that statisticians are narrow-minded or that statistical solutions do not make sense. I only contend that professionals tend to suggest solutions that are right up their alley. Teachers probably suggest teaching more statistics and university administrators sending researchers to courses in ethics, integrity, and improved PhD supervision. All these measures contribute to improving the quality of scientific activity but fail to address some of the key features of scientific research. These features are the methodologically sound design and execution of studies and the correct statistical analysis of data, all of which ideally take place in complete openness. I already discussed preregistration of planned research as part of an open research culture.

In this chapter, I discuss statistical innovation suggested as a means of improving statistical data analysis with an eye toward the possible reduction of QRPs and downplay possible optimism about this course of action. I also comment on the suggestion to teach more statistics and attempts to reduce situational influences on QRPs such as performance pressure and publication bias. Finally, I suggest actions that I consider most effective and efficient. These are the publication of data and research details, including preregistration, and seeking statistical consultation.

SUGGESTED MEASURES THAT MAY NOT BE EFFECTIVE

Another p-Value?

In recent years, many statisticians have suggested that the p-value of a statistical test, expressing the probability of a result at least as extreme as the result found from the data provided the null hypothesis is true, is the culprit of much that is wrong in how researchers use statistics. In previous chapters, I discussed p-hacking, the practice of data analysis aimed at optimizing conditions to decrease p-values, hence increase the probability of finding significant results (Ulrich & Miller, 2020; Wicherts et al., 2016). Benjamin, Berger, Johannesson, Nosek, Wagenmakers et al. (2017) did not so much focus on p-hacking but claimed that the widespread use of significance level $\alpha = .05$ sets the standards for new discoveries too low and suggested using the stricter $\alpha = .005$ instead. The smaller significance level supports a much higher Bayes factor—discussed in the next section—in favor of the alternative hypothesis in many studies. A smaller significance level also reduces the false positive rate, here meaning reduction of the proportion of rejected true null hypotheses out of the total number of rejected true and false null hypotheses. In replication research, given that the initial study used the smaller significance level, the authors claimed significant results have a higher success rate. Other undesirable effects can be controlled by keeping the test's power at $1 - \beta = .8$, which means increasing average sample size by 70%.

Responding to Benjamin et al. (2017), Lakens, Adolfi, Albers, Anvari, Apps et al. (2018) argued against one p-value for all seasons. They noticed that, of the 47 results in the Open Science Collaboration (2015) project selected for replication that had $p \leq .005$, 23 studies yielded $p \leq .05$ upon replication. This means they were "not as significant" as the original statistical testing values. I remind the reader of the counteracting regression effect discussed in Chapter 5, which predicts a result like this. Neither Benjamin et al. (2017) nor Lakens et al. (2018) discussed the regression effect as a possible and incomplete explanation of disappointing replication results. Trafimow, Amrhein, Areshenkoff, Barrera-Causil, Beh et al. (2018) did, and based on the regression effect and other influences argued against the use of p-values altogether. Rather than considering statistical argumentation about use, misuse, and uselessness of p-values, I think the recommendations of Lakens et al. beyond the use of p-values are interesting and I will come back to them when discussing suggested innovations in statistics education. Specifically, Lakens et al. recommend abolishing the *label* of statistical significance (but not statistical significance itself) and replacing it with interpretations researchers provide of their results that are more meaningful, not simply relying on one significance level. In particular, the authors suggest that researchers clarify their design choices, including (ibid., p. 169) "the alpha level, the null and alternative models, assumed prior odds, statistical power for a specified effect size of interest, the sample size, and/or the desired accuracy of estimation ... before data are collected." These and other choices should be preregistered by means of Registered Reports (Chapter 5).

Ignoring possible criticism and issues concerning preference, the recommendations of Benjamin et al. and Lakens et al. are rational and defendable from the perspectives of responsible statistics use and policy for research registration, respectively. The question in this book is whether they will reduce QRPs. If you take the insights Tversky and Kahneman (Chapter 6) provided seriously, rules of thumb such as recommended significance levels facilitate fast decision-making but discourage slow, rational thinking. If researchers take $\alpha = .05$ as their beacon, ignoring all drawbacks of which they may not even be aware, then there is little reason to expect that using $\alpha = .005$ will make them act differently. One could argue that ritually using the smaller significance level at least causes smaller damage and greater chance of successful replication, but the practice of research may be more complex (Trafimow et al., 2018). I expect that responsibly using significance levels or alternatives such as Bayes factors is the privilege of experienced statisticians and researchers equipped with enough knowledge, skills, and experience. I also expect that researchers will use different significance levels similarly, and that this will have little if any effect on the occurrence of QRPs. Finding fewer significant results using $\alpha = .005$ may even produce greater disappointment, fueling the search for alternative research strategies producing misleading results, whatever they may be (Amrhein & Greenland, 2018).

Again, I emphasize that I will not argue with the usefulness of the recommendation Lakens et al. made in the context of preregistration from a rational perspective. I will rather assess the feasibility of their proposal in the practice of research with an eye toward reduction of QRPs. Then my point is that, theoretically, if most researchers are able to make all the thought-out and competent design and statistical choices Lakens et al. suggest, one may ask why researchers do not make these choices routinely. The most likely answer is that making all these decisions correctly requires a deep understanding of statistical procedures facilitated by having learned from experience. Why do Lakens et al. introduce preregistration through Registered Reports when researchers have a sufficient understanding of statistics? Do they expect that otherwise researchers mastering statistics to the degree Lakens et al. suggest, after all prefer exploration without prior commitment, hide non-significant results in their desk drawers, and count on journals to accept significant results resulting from p-hacking? Perhaps some (or many) of them do, but we cannot know this. I assume that many researchers cannot satisfy the expectations Lakens et al. expressed and may be discouraged from preregistration when confronted with such difficult requirements and resort to the more comfortable use of blind rules of thumb, avoiding preregistration altogether.

Adopting the Bayesian Way?

Frequentist statistics is the basis of this book, and it is by far the most popular approach to statistical thinking. Bayesian statistics, which is at least as old, has seen a growing popularity in the past few decades (e.g., Van de Schoot, Kaplan, Denissen, Asendorpf, Neyer et al., 2014). A difference between the two approaches that immediately meets the eye is the absence of null-hypothesis statistical testing and the use of different criteria than the p-value for assessing hypotheses in Bayesian statistics (e.g., Gigerenzer, Kraus, & Vitouch, 2004; Wagenmakers, 2007). The reason why I will say a little bit about Bayesian statistics is that several authors (Krutschke, 2010; Wagenmakers, Marsman, Jamil, Ly, & Verhagen, 2018) expect it to play a role in reducing QRPs when researchers use it and thereby will abolish frequentist null-hypothesis statistical testing. I will discuss the core idea of Bayesian thinking and consider what it means for reducing QRPs. The interested reader can find some formal background in Appendix 7.1.

Whereas frequentist statistics follows the logic of formulating a null hypothesis one tries to reject in favor of an alternative, Bayesian statistics departs from a hypothesis of interest based on present knowledge and then assesses whether newly collected data give rise to changing the belief one has in the hypothesis. Frequentist statistics is without memory and thus tests each new null hypothesis as if it is the first time that anyone ever collected data on the topic and tests the null hypothesis. It is the researcher who must assess to what degree the new result is consistent with knowledge existing before the hypothesis testing. In contrast, Bayesian statistics tries to learn from experience by considering what the researcher knows of the topic and previously collected data and then combines this knowledge with new evidence to arrive at a modification of the state of knowledge. Statistically, Bayes' theorem (Appendix 7.1) connects old and new evidence. It is important to realize that in this case it is the researcher who must inform the procedure about her knowledge and the quality of the information supplied decides on the outcome in combination with the new evidence.

The learning principle of Bayesian statistics is that the belief you have in your hypothesis (say, hypothesis H) after you have seen the new data (denoted D), expressed as a conditional probability, $P(H|D)$, depends on your initial belief in the hypothesis before having seen the Data, $P(H)$, corrected by a factor expressing the likelihood of the Data given the Hypothesis, $P(D|H)$, relative to the likelihood of the Data without the Hypothesis (or averaged across all possible hypotheses), $P(D)$. Conditional probability $P(H|D)$ is known as the posterior probability and probability $P(H)$ is the prior probability. Bayes theorem connects these four probabilities,

$$P(H|D) = P(H) \times \frac{P(D|H)}{P(D)}.$$

Compared to your belief in the hypothesis before you did the new study, the outcome of the new study can modify your belief in either of two directions or do nothing with it. The latter situation occurs when the hypothesis does not influence the likelihood of the data and the correction factor—the fraction at the right-hand side of the equation equals 1. The data are just as likely under the hypothesis as they are without a specific hypothesis, and nothing changes. The former situations are that the data are more likely under the hypothesis, increasing your belief in the hypothesis, and that the data are less likely under the hypothesis, decreasing your belief in the hypothesis. In these two cases, the new study has changed your belief in the hypothesis.

An interesting question is what to make of the posterior probability. When it equals the prior probability, the data neither increased nor decreased your faith in the hypothesis. However, data will always influence the prior belief at least a little—why precisely nothing?—and the unavoidable question is when a change is worthwhile and when the posterior probability is of interest. And when do you have enough certainty about a hypothesis to stop initiating new studies? Statistics cannot tell you this. It can only summarize, not interpret. Interpretation is the privilege as much as the burden of the researcher. This is where researchers' degrees of freedom enter the stage forcefully: When the process no longer is objective and subjectivity takes over, anything can happen.

Bayesian statistics has the Bayes factor as an alternative for the frequentist p-value. The Bayes factor is a ratio of two probabilities concerning two competing hypotheses, H_1 and H_2, for the same phenomenon. Because it is a ratio of two probabilities, the Bayes factor, denoted BF, ranges from 0 to ∞. A value $BF = 1$ shows that the two hypotheses, H_1 and H_2, predict the data equally well and that the data make no difference for the plausibility of the two hypotheses. It is important to notice that the Bayes factor compares hypotheses, but that this means nothing about their absolute truth. Both hypotheses can be wrong. The Bayes factor shows their relative predictive performance, not their truth.

How is this different from frequentist null-hypothesis testing? You could argue that in the latter case you also compare two hypotheses, the null and the alternative hypotheses. True, but no one believes in the null hypothesis—why would two means be precisely equal?—whereas the alternative hypothesis, which is the complement of the implausible null hypothesis, is therefore rather uninformative. Aside from such considerations, even though it is quite an arbitrary decision criterion, the significance level helps the researcher to decide about the p-value, and similarly, rules of thumb exist for interpreting the Bayes factor. Suppose $BF > 1$ means that the credibility of H_1 has grown relative to the credibility of H_2, and $BF < 1$ means that its relative credibility has shrunk. Rules Kass and Raftery (1995) proposed are popular among researchers: $1 < BF \leq 3$: not worth more than a bare mention; $3 < BF \leq 10$: substantial; $10 < BF \leq 100$: strong; $BF > 100$: decisive. If the effects move in the other direction, then intervals on the [0, 1] scale are: $\frac{1}{3} \leq BF < 1$, $\frac{1}{10} \leq BF < \frac{1}{3}$, $\frac{1}{100} \leq BF < \frac{1}{10}$, and

$0 < BF < \dfrac{1}{100}$, and have the same interpretation as the other intervals. Like the frequentist literature with respect to α-values, the Bayesian literature (e.g., Van Doorn, Van den Bergh, Böhm, Dablander, Derks et al., 2021) does not show unanimity as to which guidelines to use, but the point I wish to make is that guidelines help the researcher, who would otherwise be at a loss what to make of her results (Van Belle, 2008).

This modest discussion of Bayesian statistics calls attention to an alternative approach to assessing research results, but my purpose is not to advise researchers to use Bayesian rather than frequentist statistics or vice versa. Given that there always was a controversy between frequentists and Bayesians, it is more interesting that some theorists seem to expect Bayesian statistics will reduce QRPs. However, proponents of Bayesian statistics are not blind to the human factor in statistical data analysis. For example, Wagenmakers et al. (2018) noticed that Bayesian inference does not protect against malice and statistical misunderstanding. Krutschke (2021) noticed that application of Bayesian statistics suffers from the same vulnerability as frequentist statistics when researchers cull out their inconvenient results to produce biased results, a practice Konijn, Van de Schoot, Winter, and Ferguson (2015) refer to as *BF*-hacking. However, Konijn et al. expect using Bayesian statistics to reduce the occurrence of publication bias, but Simmons, Nelson, and Simonsohn (2011) expect its use to increase researchers' degrees of freedom, increasing the likelihood of incorrect results.

The question relevant in this book is whether using Bayesian statistics prevents QRPs. Several Bayesians recognize that Bayesian statistics is not a panacea for QRP reduction. I agree and refer to the work of Tversky and Kahneman. Their System 1 deals with Bayesian and frequentist statistics just the same; that is, people confronted with an even slightly difficult problem solve it by substitution with an easier but different problem and come up with a coherent answer that, however, does not solve the initial, difficult problem. It takes experience to recognize that the difficult problem is not solvable just like that and needs much effort involving patience and perseverance. Many introductory texts promote Bayesian thinking with great enthusiasm, but some seem to underestimate the high level of rational reasoning involved, automatically pushing practical solutions into the realm of System 1. I emphasize that this is no different for rational reasoning about frequentist statistics. Both approaches are in the same boat.

More Education?

Several authors have emphasized the need of better educating students in statistical skills (e.g., Garfield, 1995; Greenhouse & Seltman, 2018; Lovett & Greenhouse, 2000; Moore, 1997; Wild & Pfannkuch, 1999). Gal (2002) and Johannssen, Chukhrova, Schmal, and Stabenow (2021) addressed the statistical literacy problem in the broader context of the public's and the media's understanding of statistics in the data science era. Inspired by the Stapel case, Asendorpf, Conner, De Fruyt, De Houwer, Denissen et al. (2013) recommended three main topics for improving students' statistical skills. First, statistics courses need to emphasize doing research that is methodologically sound rather than publishable. Second, courses need to teach statistical concepts necessary to facilitate replicable research, such as power analysis and effect size in relation to standard error, and further illuminate the value of non-significant findings and the difficulty of realizing replications. Teachers need to de-emphasize simple null-hypothesis significance testing based on single studies without replication. Moreover, courses need to encourage transparency, such as realized by means of archiving of data and analysis scripts, and let students conduct their own replication studies in an experimental context. Third, statistics courses should encourage critical thinking by training students to ask critical questions about published articles and make room for articles that discuss replicated studies and studies that were not replicable. Focusing on multiple studies should clarify concepts such as statistical power and sampling theory. Teachers should illuminate the deteriorating effects on the validity of statistical results of repeated testing while enlarging the sample in between testing until significance, data fishing, and leaving out cases based on invalid reasons, as well as the tricks leading to p-hacking. Who will argue with such great advice?

My problem is that the advice assumes that students already master a high level of reasoning based on a broad repertoire of knowledge of statistical methods and procedures, as well as an extensive experience in using statistics in data analysis. However, students do not master this large repertoire, nor do they have experience using statistics beyond a few examples. Higher-level reasoning leaning on an extensive knowledge base and a long-standing experience analyzing data is beyond their status quo as novices—except for the rare talent who needs little information to make much progress. Unless education programs make much more room for statistics courses, students will be acquainted with only basic statistical methods, such as t-tests, chi-squared tests, and nonparametric tests, essentials of linear regression models, and analysis of variance, and catch a glimpse of more advanced methods, such as factor analysis, structural equation models, and multilevel models. Their limited knowledge base will allow them to recognize popular methods when they need them, but not to reason about difficult statistical issues that require so much more knowledge and experience.

A student who came to me after a class I taught about methods for estimating agreement between judges and who apparently was highly disinterested in the topic, said to me that he preferred skipping statistics to follow clinical courses and that he would find me when he needed my advice. I replied that I expected he would not do that, given that (in his world) he never learned that judgment agreement was a problem anyway; so, how could he ever ask the question? You may think of this student whatever you like, but he simply expressed—perhaps somewhat inconsiderately—what more students were thinking. A plea from me to consider statistics training a great exercise in logical reasoning, a skill also needed when diagnosing patients, never convinced students completely to suffer the statistics course for their own good, but they did. They were probably not the future researchers this book addresses, but the problem of difficulty and experience is clear. Take the use of *p*-values I discussed: If statisticians cannot agree about a significance level, how to use it, or whether to use it at all (e.g., Greenland, Senn, Rothman, Carlin, Poole et al., 2016; Hand, 2022; McShane & Gal, 2017), what can you expect of a graduate student?

Like suggestions about the best significance level, suggestions to improve statistics education suffer from the problem that they assume an exceptionally high level of knowledge and experience researchers and students for which the methods and teaching are intended usually do not have. This does not mean that we must stop thinking about improved statistical methods and procedures, and topics to teach; of course, we need to do all of this because it improves the quality of our knowledge and advances our thinking. The crucial question is here whether the suggested topics for teaching reduce QRPs. Given that almost all programs in biology, education, health, medicine, psychology, and sociology offer a limited number of courses in basic research methods and statistics, the most one can expect is that students have some novices' knowledge and can apply some of the simpler statistical methods and procedures. It is too optimistic to expect students to be able to reason at a level that assumes they master most of the statistical methods and procedures at a more complex level. They have not had the opportunity to gain experience, which accumulates across time.

My reservations do not imply that teachers should not make students aware of what can go wrong in research and how to prevent this. When they know that research comes with imperfections and flaws, students are at least aware that research is not perfect and they, when researchers, are liable to error. Quintana (2021) proposed students writing theses based on research they did as part of their undergraduate training program perform studies replicating interesting, published studies tailored to their level of expertise and the resources the program can afford. The student learns to do research pursuing interesting questions, meanwhile experiencing the value of preregistration, replication, and sound methodology applied to interesting questions. Science, in the example psychology, profits from vastly accumulating replication evidence for relevant issues already present in the literature. The idea of experiencing and internalizing important research concepts is also present in attempts (Pennock & O'Rourke, 2017) to internalize RCR, the responsible conduct of research (Chapter 2) and take away its somewhat unappealing presence caused by a legalistic hence obligatory image. I applaud educational innovations that steer students in the right direction, reducing QRPs in later work, but note that my previous reservations remain.

Most students will not become researchers but those who do are not suddenly experienced statisticians. Add this to the discussion of practicing statistics *on the side* next to one's true specialization, and QRPs are almost a logical consequence. Once they are researchers, some people attend courses in advanced statistical methods they expect they need in their research. Such courses often are short, taking one or a few days, trying to make efficient use of researchers' time. Courses often aim higher than researchers' true levels of statistical expertise and experience allow and focus on software use more than on an understanding of the details of the methods. One could argue that, like driving a car, it suffices to know how to operate the dashboard and ignore what goes on underneath the hood. However, questionable driving practices do not depend on the level of technical knowledge about automobile engines, whereas QRPs do depend on the unintended abuse of statistical methods and procedures. A deeper understanding fed by experience seems to be mandatory.

Reducing Situational Influences on QRPs?

In several places in this book, I have mentioned situational influences that facilitate the occurrence of QRPs and provided references where you can find lists of QRPs. I also noticed that, with a few exceptions, causes of QRPs identified in the literature are mostly situational, such as employer's performance pressure and journals' publication bias, but that researchers' poor mastery of statistics is rarely mentioned as a cause of QRPs. The work of Tversky and Kahneman explains why statistical questions, not only the most difficult questions, pose insurmountable problems to many researchers and often to statisticians as well. Other authors have also noticed students' and researchers' struggle and produced an impressive number of books reaching out to readers having trouble with probability and statistics (e.g., Abelson, 1995; Hand, 2008; Rowntree, 2018; Wainer, 2016). Their common point of departure is that with the right education, training, and motivation, students, beginning researchers but also interested non-researchers can attain a deeper understanding of the principles of statistical thinking. This would help everyone understanding reports and newspaper articles using results of statistical analyses and researchers doing their data analyses correctly. Again, as with the call for better statistics education discussed in the previous section, I find it difficult to argue with this optimistic point of view and will not try to discourage anyone pursuing this path. But it is not enough.

Given that statistics is difficult and results often counter-intuitive, that social, behavioral, health, and other data are often complex containing multiple variables with high-dimensional structures that are noisy and contain many inexplicable signals, and that researchers lack experience in statistical thinking, I find it difficult to believe that QRPs can be avoided on a large scale. On the contrary, I think that the situation promotes, without intention of course, the occurrence of QRPs. In addition, researchers often consider statistics a supporting discipline for biology, medicine, psychology, and sociology, and even though this is true, it does not make statistics easier for them. On the contrary, it makes statistics more difficult, because when it becomes second, researchers more likely will spend too little time—time they need for practicing their main discipline at the highest level—accumulating the experience needed to practice statistics well. Many other disciplines have their own supporting disciplines. For example, in the nineteenth century, physics became a supporting discipline for chemistry, developing the subdiscipline of physical chemistry (Van den Berg, 2021), but I guess no one will argue that this makes physics easier to understand and practice for chemists. Similarly, chemistry became a supporting discipline for biology, and statistics, a subdiscipline of mathematics, became a supporting discipline for all disciplines collecting empirical data. On the one hand, for a statistician it is great to see statistics become so important, but on the other hand, its use by inexperienced non-experts is a cause for concern.

The many situational causes such as performance pressure imposed by an employer, the researcher's discipline, or the researcher herself, the file drawer problem, publication bias, and many other causes amplify the occurrence of QRPs. The final take-home message of Chapter 6 was:

> Query: Imagine a world in which researchers master statistics and data analysis well. Will QRPs be a problem?

My hypothesis is: No, or hardly. Of course, the take-home messages are intended for discussion, so you may disagree. My viewpoint does not imply that statisticians are people without vices; that is, if you are a statistician, it is guaranteed no data problems will occur. I expect (and know) also statisticians will make mistakes. The viewpoint does imply, however, that once you understand that the effect of various unjustified manipulations of the data or of certain incorrect uses of statistical methods is to introduce systematic and random errors that dominate and invalidate the results of the analysis, it becomes unattractive to engage in these practices. What is the fun or the use of doing something incorrectly if you know you are doing it incorrectly? Unless, of course, you *intend* to fake your results in the interest of some benefit that does not concern the empirical truth. Intentional action to fake results qualifies as FFP, as I explained in Chapter 2, even if this does not involve fabrication or falsification of data. However, it would involve falsification of *results*, hence my reference to FFP.

Assuming as before that QRPs are the result of too little knowledge and experience using statistics adequately, QRPs happen, and I expect situational influences will amplify their frequency and seriousness (IJzerman, Lewis Jr., Przybylski, Weinstein, DeBruine et al., 2020). Researchers do not like to report failures, and a non-significant result or a clearly misfitting model look like failures and may disappear in the file drawer. Journals do not like to report "failures" as well and tend to publish mainly studies that report rejected null hypotheses and fitting models, a tendency that promotes publication bias. Researchers know this and thus submit primarily studies that present rejected null hypotheses and fitting models. The circle thus seems round. Other circles amplify the bias culture. For example, employers want successful employees that contribute to the firm's mission and goals. Thus, universities hire and promote academic staff that publish in top journals that tend to publish rejected null hypotheses and fitting statistical models. In addition, grant organizations provide grants to researchers that are successful, and they happen to be researchers who publish about rejected null hypotheses and fitting models in top journals. Scientists strive for success, defined as large numbers of articles in preferably top journals, large numbers of successful supervised PhD theses, and large amounts of grant money (Miedema, 2022). They all know this and are one another's competitors for limited journal space and grant money (Fang & Casadevall, 2015; also, Aksnes, Langfeldt, & Wouters, 2019).

The competition for scarce resources and high production—articles, PhD projects, grant money—puts a lot of pressure on the scientific community. An important question is whether competition affects every researcher similarly. A culture that fosters competition may not be that bad given that competition can drive competitors to attain great results, serving the interest of society in which case everybody wins. In addition, one may ask whether science is not inherently competitive, with researchers trying to contribute to the solution of the same problems and wondering whether someone else will come up sooner with a solution. Moreover, not everybody dislikes competition, and many researchers engage in competition not paying too much attention to it and playing the game of science by the rules. The truth is not black and white but rather gray, as it is so often, but it is crucial to realize that competition may also foster QRPs when the competitors are forced to use statistics when they lack sufficient knowledge and experience to do this correctly. Competition will probably stay, and the question is how to reduce QRPs alternatively (Sijtsma, 2016).

TWO RECOMMENDATIONS

Open Data and Research Details

Researchers publish their work to let their colleagues know the results. If they would not publish, their work would reduce to a hobby that they can keep to themselves. To publish means: "to make available to the public, usually by printing, a book, magazine, newspaper, or other document."[1] One could limit this definition to the article or book chapter in which the researcher reports her results and the theoretical considerations, hypotheses, and methods that laid the foundation for the data collection and the data analysis. A broader interpretation would include all details colleagues need to check the results reported or prepare a replication study. This interpretation of publishing extends passively reporting according to the publication rules of the research area and the journal or publisher to providing all the materials colleagues need to redo the analysis or replicate the research. Redoing the data analysis can suggest alternative analysis strategies, point out weak or questionable analysis choices, and expose errors. While some researchers may prefer to withhold research details fearing the scorn of their colleagues even when only accidental errors are exposed, the upside is knowing that their way of working in an open research climate will protect them from engaging in QRPs. Knowing colleagues look across your shoulder supports working cautiously and producing the best work within your capabilities.

While an open research climate seems self-evident and necessary (e.g., Haven, Tijdink, Martinson, & Bouter, 2019; Nosek, Alter, Banks, Borsboom, Bowman et al., 2015; Simonsohn, 2013; Wicherts & Bakker, 2012), few researchers publish their data and other research details colleagues need for checking results or replicating a study (e.g., Hardwicke & Ioannidis, 2018; Vines, Albert, Andrew, Débarre, Bock et al., 2014, Wallach, Boyack, & Ioannidis, 2018; Wicherts, Borsboom, Kats, & Molenaar, 2006). Possible reasons for this omission are the following.

- *Loss of Investment.* Collecting data costs money, personnel, and effort. Data are the fuel that makes the researcher's engine run. Without data, researchers cannot contribute to the progress of their research area and threaten to lose their purpose. So, why give your data to others just like that? Why run the risk that others scoop you and publish new results using your data that you did not think of yourself or did not have the time yet to publish in an additional article? Let them collect their own data! Much as this attitude is understandable, it flies in the face of the general interest science has in being credible, and credibility profits from an open research culture. Although I cannot know this with certainty, I guess that much of Stapel's misconduct might never have occurred if he had known that all he did would be out on the Web for everybody to see and reveal.

A helping hand to accommodate researchers' worries is to allow them to publish their data and other research details directly after they have published their first (second, third, et cetera) article and have possible additional papers in preparation. This gives them a head start compared to colleagues who are looking for easy data rather than redoing the analysis or replicating research, which are obviously of common interest. Other measures protecting researchers' interests do not seem complex, but I will refrain from further working them out.

- *Ownership.* A third party rather than the university or another employer may be the owner of the data. Examples are government organizations, municipalities, commercial enterprises, and hospitals. With medical research, pharmaceutical companies may be the owners, and with psychological and health research, nursing homes may provide data they claim they own. If the researcher can simply publish her results along with the research details and the data, and other researchers have full access to all materials, there is no problem. However, when third parties facilitating research limit or hinder making research details and data publicly accessible and usable, the key principle of open science is at risk.

  Solutions are that researchers (or their employers) negotiate with third parties about the possibility to publish the results of their study without limitation, as well as the publication of the data and relevant research details. A reason why third parties might hesitate or refuse is that they fear public knowledge about the organization may damage their interest. Instead of a university or another organization claiming to do scientific research, third parties can hire another research partner not bound to openness to cooperate with or do their research. The research results will then be used only in the organization's interest but will not contribute to the publicly available body of knowledge science represents. The prospect of generous funding may be hard to resist for academic researchers and university administrators, moving them into a conflict between research ethics and financial or other interests.

Clearly, an open-data culture raises additional problems, and we need experience to develop a practice that works. Ioannidis (2016) noticed that making errors is unavoidable and discussed incentivizing researchers to publish their data without fear of being ridiculed when errors are exposed or results are not replicated. He also noted that there is no guarantee that authors will publish the data they used and that researchers re-analyzing data are skilled and do not have interests other than intellectual ones. To reduce some of the problems, Ioannides suggested publishing information about design and analysis details in addition to the data and pre-registering the study. I add that refraining from an open-data policy means choosing to rely blindly on whatever a colleague reports in an article giving up the possibility of control. This places the research outside the realm of science.

Finally, a measure I rarely encountered in the literature on QRPs and related issues is employers such as university executive boards requiring researchers—the university's employees—to sign for open-data policy as part of their employment contract. I am not saying force gets things done automatically, but it would help tremendously having open-data policy accepted when employees know their institution considers publishing data self-evident and facilitates doing this effectively and efficiently.

In der Beschränkung zeigt sich erst der Meister: Statistical Consultancy

The literature on research integrity and QRPs is replete with advice for researchers about improved statistical procedures, but nowhere did I read the simplest advice: *Leave it to an expert.*[2] Some researchers are excellent statisticians, but other researchers practice statistics in the side, next to their focus area of expertise. Statistics is a vast academic discipline and known to be quite difficult, so thinking you can manage yourself is a remarkable choice when you are not a statistics expert. Why do many researchers not ask a statistician's advice when faced with an awkward data-analysis problem? They may be unaware of the difficulty of statistics and their unconscious System 1 response, rely on advice given by an inexperienced colleague, or overrate their degree of experience and their skills in statistics. And when needed, a statistician may simply be unavailable because the institute did not hire statisticians or because they are too busy doing other things.

A reviewer of this book suggested that statisticians might be embedded in psychology, health, medical, and other research groups to optimize the match between the group's needs for specific statistical models and the statistician's expertise. I agree this is an ideal situation, but I will explain why it is unrealistic in many cases. First, like psychology, health studies, and medicine, statistics is a large and diverse discipline, and statisticians need time to be trained in the intricacies of their discipline. As an academic discipline, statistics is no different from the other disciplines. Statisticians affiliated with universities teach their statistics courses to statistics undergraduates, graduates, and broader student audiences. They also organize their research developing new statistical methods and software in statistics research groups. By doing this, they increase their skills and experience and become better statisticians. If statisticians were embedded in research groups in other disciplines, the development and improvement of statistical methods and software probably would no longer be a priority. This would be at the detriment of their skills and reduce their ability to make the right statistical choices. Like any academic, statisticians need their own habitat to develop and grow as scientists.

To help other researchers with their methodological and statistical decisions, a part of statisticians' time needs to be devoted to consultation. Statisticians who are formally included in a research group usually also have a formal appointment in a statistics group to stay involved in statistics teaching and research and further develop their skills and experience. Working together with researchers from other disciplines is known to be fruitful for statisticians as well, and a breeding ground of novel ideas for further statistical development. There is already much cooperation between researchers from substantive disciplines and statisticians, but the cooperation needs to be intensified to increase the responsible use of the difficult and counterintuitive statistical methods. I expect this to reduce QRPs. Intensifying cooperation may be limited for two reasons. First, the number of statisticians, including methodologists and applied statisticians, is limited and so is the volume of expertise available for consultation. Second, what helps best is that researchers from other disciplines recognize their need for skillful statistical advice fed by ample experience. Statistics exercised on the side is a continuing risk for RCR, responsible conduct of research (Chapter 2).

The famous mathematician and statistician John Tukey is often quoted for his observation that "the best thing about being a statistician is that you get to play in everyone's backyard."[3] I do not know whether by emphasizing *playing* rather than living in other people's backyard he also had in mind that a statistician functions best when she can maintain her independence giving statistical advice. Independence guarantees she can be open and honest without suffering from possible pressure for advice clients prefer or consequences of unwanted advice. In addition to fostering professional expertise in statistics groups, independence is the second reason why I believe that statisticians function best when they can advise other researchers.

Responding to the replication crisis in several research areas, Romero and Sprenger (2020) discussed three types of reform for addressing the replication crisis, the social (e.g., more education), the methodological (e.g., preregistration), and the statistical (e.g., using Bayesian statistics), but did not mention statistical consultation. Whereas a toothache readily motivates one to see a dentist and a leaking roof to see a plumber, perhaps recommending seeing a statistician feels like defeat, assuming researchers prefer to solve their own problems and statisticians know this, rendering them cautious. I do not rule out the possibility that a statistician is a little cautious too not to invite everybody to come to her office for advice.

When I first taught a statistics course for some 150 undergraduate students, more than anything else I sensed their fear of math, an experience common to every statistics teacher. I tried to comfort them by telling them that statistics was not that difficult, and that hard work, perseverance, and patience would prove me right when they would take the test. However, this only worked until the next topic came along that scared them stiff, and many topics had this effect. After this ritual repeated for a couple of years, it finally dawned on me that thousands of students could not all be wrong: *Statistics was difficult!* From that moment, I was more cautious telling them not to be scared when they felt scared. To graduate students, I simply told the truth—statistics is difficult, but let us see how far we can get. To researchers, I say: There is no weakness or shame in not knowing things that are not your expertise. Admitting this is the challenge, and if you do, asking a statistician for help is the obvious course of action.

TAKE-HOME MESSAGES

Statistical measures such as reducing significance levels and adopting Bayesian data analysis methodology will not reduce QRPs resulting from insufficient experience using statistical techniques.

Improving statistics education usually amounts to asking more of students who have neither the knowledge nor the experience available at their young age.

Considering full transparency about research design and data, including preregistration of research plans, and asking for statistical consult when needed are the most effective means of preventing QRPs.

APPENDIX 7.1: BAYESIAN STATISTICS

Bayesian statistics rests on the principle that the belief you have in your hypothesis H after you have seen the new data denoted D, and expressed as a probability, $P(H|D)$, depends on your initial belief before having seen the Data, $P(H)$, corrected by a factor expressing the likelihood of the Data given the Hypothesis, $P(D|H)$, relative to the likelihood of the Data without the Hypothesis, $P(D)$. The four probabilities are connected by the next equation,

$$P(H \mid D) = P(H) \times \frac{P(D \mid H)}{P(D)}.$$

Probability $P(H|D)$ expresses the belief in the Hypothesis given the Data, hence, called the posterior (i.e., subsequent, following knowledge of the new data) probability. It is obtained from the probability of the Hypothesis before having seen the Data, $P(H)$, called the prior (meaning, preceding the new data) probability.

Prior probability $P(H)$ transforms to posterior probability $P(H|D)$ by multiplication with $\dfrac{P(D \mid H)}{P(D)}$. This ratio is the predictive updating factor. It updates your belief by comparing the probability of the newly collected Data given the Hypothesis (numerator) with the probability of the Data averaged across all hypotheses (denominator). If the data are more likely if the Hypothesis is true than without a hypothesis, the ratio exceeds 1 and increases one's belief in the Hypothesis; that is, posterior probability $P(H|D)$ is larger, expressing a stronger belief in H than prior probability $P(H)$ did. The opposite conclusion follows from a ratio smaller than 1.

The Bayesian analogue of the frequentist p-value is the Bayes factor. For two competing hypotheses H_1 and H_2, the Bayes factor (BF) is obtained from assessing the ratio of the preceding equations for H_1 and H_2, resulting in

$$\frac{P(H_1 \mid D)}{P(H_2 \mid D)} = \frac{P(H_1) \times \dfrac{P(D \mid H_1)}{P(D)}}{P(H_2) \times \dfrac{P(D \mid H_2)}{P(D)}} \Leftrightarrow \frac{P(H_1 \mid D)}{P(H_2 \mid D)} = BF \times \frac{P(H_1)}{P(H_2)},$$

in which

$$BF = \frac{P(D \mid H_1)}{P(D \mid H_2)}.$$

The second ratio on the right of the double arrow, $\dfrac{P(H_1)}{P(H_2)}$, compares the two prior probabilities of the two hypotheses before seeing the data and is called the prior odds of the two hypotheses. The first ratio, $\dfrac{P(H_1 \mid D)}{P(H_2 \mid D)}$, compares the two posterior probabilities of the two hypotheses after seeing the data and is called the posterior odds. The Bayes factor compares the predictive performance of two hypotheses and shows how the degree to which the data transform the prior odds into the posterior odds. Because the Bayes factor is the ratio of two probabilities, it ranges from 0 to ∞. A value $BF = 1$ shows that the two hypotheses predict the data equally well and that the data make no difference for the plausibility of the two hypotheses; the posterior odds equal the prior odds. It is important to notice that I am only comparing hypotheses, which means nothing about their absolute truth. Both hypotheses can be wrong. The Bayes factor shows their relative predictive performance, not their truth.

NOTES

1 https://dictionary.cambridge.org/dictionary/english/publish

2 Section title based on Johann Wolfgang von Goethe, from his sonnet "Natur und Kunst" (1802). Translation: A master craftsman is able to restrict himself.

3 I am not aware of a reference, but the quote is nevertheless famous among statisticians.

References

Abelson, R. P. (1995). *Statistics as principled argument*. Hilsdale, NJ: Erlbaum.

Aksnes, D. W., Langfeldt, L., & Wouters, P. (2019). Citations, citation indicators, and research quality: An overview of basic concepts and theories. *SAGE Open*. doi:10.1177/2158244019829575

Allison, P. D. (2002). *Missing data. Series: Quantitative applications in the social sciences*. Thousand Oaks, CA: Sage.

Al-Marzouki, S., Evans, S., Marshall, T., & Roberts, I. (2005). Are these data real? Statistical methods for the detection of data fabrication in clinical trials. *BMJ (Clinical research ed.), 331* (7511), 267–270. https://doi.org/10.1136/bmj.331.7511.267

Amrhein, V., & Greenland, S. (2018). Remove, rather than redefine, statistical significance. *Nature Human Behavior*. doi:10.1038/s41562-017-0224-0

Anderson, M. S., Ronning, E. A., De Vries, R., & Martinson, B. C. (2007). The perverse effects of competition on scientists' work and relationships. *Science and Engineering Ethics, 13*, 437–461. https://doi.org/10.1007/s11948-007-9042-5

Arkes, H. R. (2008). Being an advocate for linear models of judgment is not an easy life. In J. I. Krueger (Ed.), *Rationality and social responsibility: Essays in honor of Robyn Mason Dawes* (pp. 47–70). New York: Psychology Press.

Asendorpf, J. B., Conner, M., De Fruyt, F., De Houwer, J., Denissen, J. J. A. et al. (2013). Recommendations for increasing replicability in psychology. *European Journal of Personality, 27*, 108–119.

Azevedo, C. D. S., Gonçalves, R. F., Gava, V. L., & Spinola, M. D. M. (2021). A Benford's Law based methodology for fraud detection in social welfare programs: Bolsa Familia analysis. *Physica A: Statistical Mechanics and its Applications*. https://doi.org/10.1016/j.physa.2020.125626

Baker, M. (2016). Is there a reproducibility crisis? *Nature, 533*, 452–454.

Bakker, M., Veldkamp, C. L. S., Van Assen, M. A. L. M., Crompvoets, E. A. V., Ong, H. H. et al. (2020). Ensuring the quality and specificity of preregistrations. *PLoS Biol, 18*(12), e3000937. https://doi.org/10.1371/journal.pbio.3000937

Barnett, V., & Lewis, T. (1994). *Outliers in statistical data.* Chichester, UK: Wiley.

Baud, M., Legêne, S., & Pels, P. (2013). *Circumventing reality. Report on the anthropological work of professor emeritus M. M. G. Bax.* Downloaded from: https://www.vu.nl/en/Images/20131112_Rapport_Commissie_Baud_Engelse_versie_definitief_tcm270-365093.pdf

Benford, F. (1938). The law of anomalous numbers. *Proceedings of the American Philosophical Society, 78*, 551–572.

Benjamin, D. J., Berger, J. O., Johannesson, M., Nosek, B. A., Wagenmakers, E.-J. et al. (2017). Redefine statistical significance. *Nature Human Behavior.* https://doi.org/10.1038/s41562-017-0189-z Downloaded from: https://www.nature.com/articles/s41562-017-0189-z

Benjamini, Y., & Hochberg, Y. (1995). Controlling the false discovery rate: A practical and powerful approach to multiple testing. *Journal of the Royal Statistical Society, Series B, 57*, 289–300.

Borsboom, D., Van der Maas, H., Dalege, J., Kievit, R., & Haig, B. (2021). Theory construction methodology: A practical framework for theory formation in psychology. *Perspectives on Psychological Science, 16*(4), 756–766. https://doi.org/10.1177/1745691620969647

Bos, J. (2020). *Research ethics for students in the social sciences.* Cham: Springer.

Bouri, S., Shun-Shin, M. J., Cole, G. D., Mayet, J., & Francis, D. P. (2014). Meta-analysis of secure randomized controlled trials of β-blockade to prevent perioperative death in non-cardiac surgery. *Heart, 100*, 456–464. https://doi.org/10.1136/heartjnl-2013-304262

Bouter, L. M., Tijdink, J., Axelsen, N., Martinson, B. C., & Ter Riet, G. (2016). Ranking major and minor research misbehaviors: Results from a survey among participants of four World Conferences on Research Integrity. *Research Integrity and Peer Review, 1*, 17. https://doi.org/10.1186/s41073-016-0024-5

Broad, W., & Wade, N. (1982). *Betrayers of the truth: Fraud and deceit in the halls of science.* New York: Simon and Schuster.

Buck, H. M., Koole, L. H., Van Genderen, M. H. P., Smit, L., Geelen, J. L. M. C. et al. (1990). Phosphate-methylated DNA aimed at HIV-1 RNA loops and integrated DNA inhibits viral infectivity. *Science, 248*(4952), 208–212. https://doi.org/10.1126/science.2326635

Budd, J. M. (2013). The Stapel Case: An object lesson in research integrity and its lapses. *Synesis. A Journal of Science, Technology, Ethics, and Policy.* Downloaded from https://englishdocs.eu/wp-content/uploads/2018/06/budd_2013_g47-53.pdf

Callaway, E. (2011). Report finds massive fraud at Dutch universities. *Nature.* Nov 1; 479 (7371): 15. https://doi.org/10.1038/479015a

Campbell, S. K. (1974, 2002). *Flaws and fallacies in statistical thinking.* Mineola: Dover Publications, Inc.

Chambers, C. D. (2013). Registered reports: A new publishing initiative at Cortex. *Cortex, 49,* 609–610. http://dx.doi.org/10.1016/j.cortex.2012.12.016

Chambers, C. D. (2017). *The 7 deadly sins of psychology: A manifesto for reforming the culture of scientific practice.* Princeton, NJ: Princeton University Press.

Chevassus-au-Louis, N. (2019). *Fraud in the lab. The high stakes of scientific research.* Cambridge, MA: Harvard University Press.

Cohen, J. (1960, 1988). *Statistical power analysis for the behavioral sciences.* Hillsdale, NJ: Erlbaum.

Cole, G. D., & Francis, D. P. (2014a). Perioperative β blockade: Guidelines do not reflect the problems with the evidence from the DECREASE trials. *BMJ: British Medical Journal, 349.* https://www.jstor.org/stable/10.2307/26516964?seq=1&cid=pdfreference#references_tab_contents

Cole, G. D., & Francis, D. P. (2014b). The challenge of delivering reliable science and guidelines: opportunities for all to participate. *European Heart Journal, 35,* 2435–2440.

Craig, R., Cox, A., Tourish, D., & Thorpe, A. (2020). Using retracted journal articles in psychology to understand research misconduct in the social sciences: What is to be done? *Research Policy, 49.* https://doi.org/10.1016/j.respol.2020.103930

Craig, R., Pelosi, A., & Tourish, D. (2020). Research misconduct complaints and institutional logics: The case of Hans Eysenck and the British Psychological Society. *Journal of Health Psychology, 26,* 296–311. https://doi.org/10.1177/1359105320963542

De Vries, R., Anderson, M. S., & Martinson, B. C. (2006). Normal misbehavior: Scientists talk about the ethics of research. *Journal of Empirical Research on Human Research Ethics, 1*(1), 43–50. https://doi.org/10.1525/jer.2006.1.1.51

Deckert, J., Myagkov, M., & Ordeshook, P. C. (2011). Benford's law and the detection of election fraud. *Political Analysis, 19,* 245–268. https://doi.org/10.1093/pan/mpr014

Diekmann, A. (2007). Not the first digit! Using Benford's Law to detect fraudulent scientific data. *Journal of Applied Statistics, 34*, 321–329.

Dolan, C. V. (1994). Factor analysis of variables with 2, 3, 5 and 7 response categories: A comparison of categorical variable estimators using simulated data. *British Journal of Mathematical and Statistical Psychology, 47*, 309–326.

Dyson, G. (2012). *Turing's cathedral. The origins of the digital universe.* London: Penguin Books.

Edouard, L., & Senthilselvan, A. (1997). Observer error and birthweight: Digit preference in recording. *Public Health, 111*, 77–79. Downloaded from: https://www.sciencedirect.com/science/article/pii/S0033350697900044

Epskamp, S. & Nuijten, M. B. (2016). statcheck: Extract statistics from articles and recompute p values. Retrieved from http://CRAN.R-project.org/package=statcheck. (R package version 1.2.2)

Eronen, M. I., & Bringmann, L. F. (2021). The theory crisis in psychology: How to move forward. *Perspectives on Psychological Science, 16*(4), 779–788. https://doi.org/10.1177/1745691620970586

Fanelli, D. (2009). How many scientists fabricate and falsify research? A systematic review and meta-analysis of survey data. *PLoS ONE, 4*(4), e5738. https://doi.org/10.1371/journal.pone.0005738.v

Fang, F. C., & Casadevall, A. (2015). Competitive science: Is competition ruining science? *Infection and Immunity, 83*, 1229–1233. https://doi.org/10.1128/IAI.02939-14.

Fewster, R. M. (2009). A simple explanation of Benford's Law. *The American Statistician, 63*, 26–32.

Fiedler, K., & Prager, J. (2018). The regression trap and other pitfalls of replication science—illustrated by the report of the Open Science Collaboration. *Basic and Applied Social Psychology, 40*(3), 115–124. https://doi.org/10.1080/01973533.2017.1421953

Fink, M. Gartner, J., Harms, R., & Hatak, I. (2023). Ethical orientation and research misconduct among business researchers under the condition of autonomy and competition. *Journal of Business Ethics, 183*, 619–636. https://doi.org/10.1007/s10551-022-05043-y

Fox, J. (1997). *Applied regression analysis, linear models, and related models.* Thousand Oaks, CA: Sage.

Freund, J. E. (1973, 1993). *Introduction to probability.* Mineola: Dover Publications, Inc.

Gal, I. (2002). Adults' statistical literacy: Meanings, components, responsibilities. *International Statistical Review, 70*, 1–25. https://www.jstor.org/stable/1403713

Galton, F. (1889). *Natural inheritance*. London: Macmillan.

Gardenier, J. & Resnik, D. (2002). The misuse of statistics: Concepts, tools, and a research agenda. *Accountability in Research, 9*(2), 65–74. https://doi.org/10.1080/08989620212968.

Garfield, J. (1995). How students learn statistics. *International Statistical Review, 63*, 25–34.

Gelman, A., & Hill, J. (2007). *Data analysis using regression and multilevel/hierarchical models*. Cambridge: Cambridge University Press.

Gelman, A., & Loken, E. (2014). The statistical crisis in science. *American Scientist, 102*(6), 460–465.

Gigerenzer, G., Krauss, S., & Vitouch, O. (2004). The null ritual: What you always wanted to know about significance testing but were afraid to ask. In Kaplan, D. (Ed.) *The Sage handbook of quantitative methodology for the social sciences* (pp. 391–408). Thousand Oaks, CA: Sage.

Gleick, J. (2011). *The information. A history, a theory, a flood*. New York: Vintage Books.

Goodstein, D. (2010). *On fact and fraud. Cautionary tales from the front lines of science*. Princeton, NJ: Princeton University Press.

Gopalakrishna, G., ter Riet, G., Vink, G. Stoop, I., Wicherts, J. M. et al. (2022). Prevalence of questionable research practices, research misconduct and their potential explanatory factors: A survey among academic researchers in The Netherlands. *PLoS ONE, 17*(2), e0263023. https://doi.org/10.1371/journal.pone.0263023

Greenhouse, J. B., & Seltman, H. J. (2018). On teaching statistical practice: From novice to expert. *The American Statistician, 72*, 147–154. https://doi.org/10.1080/00031305.2016.1270230

Greenland, S., Senn, S. J., Rothman, K. J., Carlin, J. B., Poole, C., et al. (2016). Statistical tests, P values, confidence intervals, and power: A guide to misinterpretations. *European Journal of Epidemiology, 31*, 337–350. https://doi.org/10.1007/s10654-016-0149-3

Grove, W. M., & Meehl, P. E. (1996). Comparative efficiency of informal (subjective, impressionistic) and formal (mechanical, algorithmic) prediction procedures: The clinical—statistical controversy. *Psychology, Public Policy, and Law, 2*, 293–323.

Hambleton, R. K., Swaminathan, H., & Rogers, H. J. (1991). *Fundamentals of item response theory*. Newbury Park, CA: Sage.

Hand, D. (2014). *The improbability principle. Why coincidences, miracles and rare events happen every day*. London: Penguin Books.

Hand, D. J. (2008). *Statistics. A very short introduction*. New York: Oxford University Press.

Hand, D. J. (2022). Trustworthiness of statistical inference. *Journal of the Royal Statistical Society: Series A (Statistics in Society), 185,* 329–347. https://doi.org/10.1111/rssa.12752

Hardwicke, T. E., & Ioannidis, J. P. A. (2018). Populating the Data Ark: An attempt to retrieve, preserve, and liberate data from the most highly-cited psychology and psychiatry articles. *PLoS ONE, 13*(8), e0201856. https://doi.org/10.1371/journal.pone.0201856

Hardwicke, T. E., Thibault, R. T., Kosie, J. E., Wallach, J. D., Kidwell, M. C. et al. (2021). Estimating the prevalence of transparency and reproducibility-related research practices in psychology (2014–2017). *Perspectives on Psychological Science.* https://doi.org/10.1177/1745691620979806

Haven, T. L., Bouter, L. M., Smulders, Y. M., & Tijdink, J. K. (2019). Perceived publication pressure in Amsterdam: Survey of all disciplinary fields and academic ranks. *PLoS One, 14*(6), e0217931. https://doi.org/10.1371/journal.pone.0217931

Haven, T. L., Tijdink, J. K., Martinson, B. C., & Bouter, L. M. (2019). Perceptions of research integrity climate differ between academic ranks and disciplinary fields: Results from a survey among academic researchers in Amsterdam. *PLOS ONE, 14*(1), e0210599. https://doi.org/10.1371/journal.pone.0210599

Haven, T. L., & Van Grootel, L. (2019). Preregistering qualitative research. *Accountability in Research.* https://doi.org/10.1080/08989621.2019.1580147

Haven, T. L., & Van Woudenberg, R. (2021). Explanations of research misconduct, and how they hang together. *Journal for General Philosophy of Science, 52,* 543–561. https://doi.org/10.1007/s10838-021-09555-5

Hays, W. L. (1994). *Statistics.* Fort Worth, TX: Harcourt Brace College Publishers.

Head, M. L., Holman, L., Lanfear, R., Kahn, A. T., & Jennions, M. D. (2015). The extent and consequences of P-hacking in science. *PLoS Biology, 13*(3), e1002106. https://doi.org/10.1371/journal.pbio.1002106

Hill, T. P. (1995). A statistical derivation of the significant-digit law. *Statistical Science, 10,* 354–363.

Huistra, P., & Paul, H. (2021). Systemic explanations of scientific misconduct: Provoked by spectacular cases of norm violation? *Journal of Academic Ethics.* https://doi.org/10.1007/s10805-020-09389-8

IBM Corp. (2021). *IBM SPSS Statistics for Windows, Version 28.0.* Armonk: IBM Corp.

IJzerman, H., Lewis Jr., N. A., Przybylski, A. K., Weinstein, N., DeBruine, L. et al. (2020). Use caution when applying behavioural science to policy. *Nature Human Behaviour, 4*, 1092–1094. Downloaded from: https://www.nature.com/articles/s41562-020-00990-w

Ioannidis, J. P. A. (2005). Why most published research findings are false. *PLoS Med, 2*(8), e124.

Ioannidis, J. P. A. (2016). Anticipating consequences of sharing raw data and code and of awarding badges for sharing. *Journal of Clinical Epidemiology, 70*, 258–260.

Ioannidis, J. P. A. (2018). Why replication has more scientific value than original discovery. *Behavioral and Brain Sciences, 41*, e137. https://doi.org/10.1017/S0140525X18000729. PMID: 31064545.

Johannssen, A., Chukhrova, N., Schmal, F., & Stabenow, K. (2021). Statistical literacy—Misuse of statistics and its consequences. *Journal of Statistics and Data Science Education.* https://doi.org/10.1080/10691898.2020.1860727

John, L. K., Loewenstein, G., & Prelec, D. (2012). Measuring the prevalence of questionable research practices with incentives for truth telling. *Psychological Science, 23*, 524–532.

Johnson, D. R., & Ecklund, E. H. (2016). Ethical ambiguity in science. *Science and Engineering Ethics, 22*(4), 989–1005. https://doi.org/10.1007/s11948-015-9682-9

Judson, H. F. (2004). *The great betrayal. Fraud in science.* Orlando, FL: Harcourt, Inc.

Kahneman, D. (2011). *Thinking, fast and slow.* London: Penguin Books.

Kahneman, D., & Tversky, A. (1982). On the study of statistical intuitions. *Cognition, 11*, 1123–1141.

Kass, R. E., & Raftery, A. E. (1995). Bayes factors. *Journal of the American Statistical Association, 90*, 773–795.

Kerr, N. L. (1998). HARKing: Hypothesizing after the results are known. *Personality and Social Psychology Review, 2*, 196–217.

Kevles, D. J. (1998). *The Baltimore case. A trial of politics, science, and character.* New York: W. W. Norton & Company, Inc.

Klaassen, C. A. J. (2018). Preliminary version. Evidential value in ANOVA-regression results in scientific integrity studies. Downloaded from: https://arxiv.org/pdf/1405.4540.pdf

KNAW; NFU; NWO; TO2-federatie; Vereniging Hogescholen; VSNU (2018). *Nederlandse gedragscode wetenschappelijke integriteit (Netherlands code of conduct for research integrity).* DANS https://doi.org/doi.org/10.17026/dans-2cj-nvwu

Konijn, E. A., Van de Schoot, R., Winter, S. D., & Ferguson, C. J. (2015). Possible solution to publication bias through Bayesian statistics, including proper null hypothesis testing. *Communication Methods and Measures, 9*(4), 280–302. https://doi.org/10.1080/19312458.2015.1096332

Krutschke, J. K. (2010). What to believe: Bayesian methods for data analysis. *Trends in Cognitive Sciences, 14,* 293–300. https://doi.org/10.1016/j.tics.2010.05.001

Krutschke, J. K. (2021). Bayesian analysis reporting guidelines. *Nature Human Behavior.* https://doi.org/10.1038/s41562-021-01177-7

Labib, K., Tijdink, J., Sijtsma, K., Bouter, L., Evans, N. et al. (2023). How to combine rules and commitment in fostering research integrity? *Accountability in Research.* https://doi.org/10.1080/08989621.2023.2191192

Lakens, D. (2019). The value of preregistration for psychological science: A conceptual analysis. *Japanese Psychological Review, 62,* 221–230.

Lakens, D., Adolfi, F. G., Albers, C. J., Anvari, F., Apps, M. A. J. et al. (2018). Justify your alpha. *Nature Human Behaviour, 2,* 168–171.

Levelt Committee, Noort Committee, Drenth Committee. (2012). *Flawed science: The fraudulent research practices of social psychologist Diederik Stapel.* Downloaded from http://www.tilburguniversity.edu/nl/nieuws-en-agenda/finalreportLevelt.pdf.

Levelt, W. A. (2011). *Interim-rapportage inzake door Prof. D.A. Stapel gemaakte inbreuk op wetenschappelijke integriteit (Interim report regarding the breach of scientific integrity committed by Prof. D.A. Stapel).* Downloaded from: https://ktwop.files.wordpress.com/2011/10/stapel-interim-rapport.pdf

Lovett, M. C., & Greenhouse, J. B. (2000). Applying cognitive theory to statistics instruction. *The American Statistician, 54,* 196–206. https://doi.org/10.1080/00031305.2000.10474545

Lüscher, T. F., Gersh, B., Landmesser, U., & Ruschitzka, F. (2014). Is the panic about beta-blockers in perioperative care justified? *European Heart Journal, 35,* 2442–2444. https://doi.org/10.1093/eurheartj/ehu056

Maddox, J. (1990). Dutch cure for AIDS is discredited. *Nature, 347,* 411.

Markowitz, D. M., & Hancock, J. T. (2014). Linguistic traces of a scientific fraud: The case of Diederik Stapel. *PLoS ONE, 9*(8), e105937. https://doi.org/10.1371/journal.pone.0105937

Masicampo, E. J., & Lalande, D. R. (2012). A peculiar preference of p values just below .05. *The Quarterly Journal of Experimental Psychology, 65,* 2271–2279.

Maxwell, S. E., & Delaney, H. D. (2004). *Designing experiments and analyzing data. A model comparison perspective.* New York: Psychology Press, Taylor & Francis Group.

McShane, B. B., & Gal, D. (2017). Statistical significance and the dichotomization of evidence (with discussion). *Journal of the American Statistical Association, 112,* 885–908. https://doi.org/10.1080/0162 1459.2017.1289846

Meehl, P. E. (1954). *Clinical versus statistical prediction: A theoretical analysis and a review of the evidence.* Minneapolis: University of Minnesota Press.

Mellenbergh, G. J. (2019). *Counteracting methodological errors in behavioral research.* Cham: Springer.

Merton, R. K. (1973). *The sociology of science. Theoretical and empirical investigations.* Chicago, IL: The University of Chicago Press.

Miedema, F. (2022). *Open science. The very idea.* Dordrecht: Springer Nature B. V.

Moore, D. S. (1997). New pedagogy and new content: The case of statistics (with discussion). *International Statistical Review, 65,* 123–165.

Nelson, L. D., Simmons, J., & Simonsohn, U. (2018). Psychology's Renaissance. *Annual Review of Psychology, 69,* 511–534. https://doi. org/10.1146/annurev-psych-122216-011836

Nietert, P. J., Wessell, A. M., Feifer, C., & Ornstein, S. M. (2006). Effect of terminal digit preference on blood pressure measurement and treatment in primary care. *American Journal of Hypertension, 19,* 147–152. Downloaded from: https://academic.oup.com/ajh/article/ 19/2/147/128590

Nosek, B. A., Alter, G., Banks, G. C., Borsboom, D., Bowman, S. D. et al. (2015). Promoting an open research culture. Author guidelines for journals could help to promote transparency, openness, and reproducibility. *Science, 348*(6242), 1422–1425. https://doi.org/10.1126/ science.aab2374

Nosek, B. A., Ebersole, C. R., DeHaven, A. C., & Mellor, D. T. (2018). The preregistration revolution. *PNAS, 115,* 11. www.pnas.org/cgi/ doi/10.1073/pnas.1708274114

Nosek, B. A., & Lakens, D. (2014). Registered reports. A method to increase the credibility of published Results. *Social Psychology, 45*(3), 137–141. https://doi.org/10.1027/1864-9335/a000192

Nosek, B. A., Spies, J. R., & Motyl, M. (2012). Scientific Utopia: II. Restructuring incentives and practices to promote truth over publishability. *Perspectives on Psychological Science, 7*(6), 615–631.

Nuijten, M. B., Hartgerink, C. H. J., Van Assen, M. A. L. M., Epskamp, S., & Wicherts, J. M. (2016). The prevalence of statistical reporting errors in psychology (1985-2013). *Behavior Research Methods,* *48*(4), 1205–1226. https://doi.org/10.3758/s13428-015-0664-2

Nuijten, M. B., Van Assen, M. A. L. M., Veldkamp, C. L. S., & Wicherts, J. M. (2015). The replication paradox: Combining studies can decrease accuracy of effect size estimates. *Review of General Psychology, 19*(2), 172–182. https://doi.org/10.1037/gpr0000034

Open Science Collaboration (2015). Estimating the reproducibility of psychological science. *Science, 349,* aac4716. https://doi.org/10.1126/science.aac4716

Pennock, R. T., & O'Rourke, M. (2017). Developing a scientific virtue-based approach to science ethics training. *Science and Engineering Ethics, 23,* 243–262. https://doi.org/10.1007/s11948-016-9757-2

Pinker, S. (2021). *Rationality: What it is, why it seems scarce, why it matters.* New York: Viking.

Pituch, K. A., & Stevens, J. P. (2016). *Applied multivariate statistics for the social sciences* (6th ed.). New York: Routledge.

Quintana, D. S. (2021). Replication studies for undergraduate theses to improve science and education. *Nature Human Behavior.* https://doi.org/10.1038/s41562-021-01192-8

Ritchie, J., Lewis, J., McNaughton Nicholls, C., & Ormston, R. (2013, Eds.). *Qualitative research practice: A guide for social science students and researchers.* London: Sage.

Romero, F., & Sprenger, J. (2020). Scientific self-correction: The Bayesian way. *Synthese.* https://doi.org/10.1007/s11229-020-02697-x

Rosenthal, R. (1979). The "file drawer problem" and tolerance for null results. *Psychological Bulletin, 86,* 638–641.

Rowntree, D. (2018). *Statistics without tears. An introduction for non-mathematicians.* Penguin Books.

Rubin, M. (2020). Does preregistration improve the credibility of research findings? *The Quantitative Methods for Psychology, 16*(4), 376–390. https://doi.org/10.20982/tqmp.16.4.p376

Salsburg, D. S. (2017). *Errors, blunders, and lies. How to tell the difference.* Boca Raton, FL: CRC Press/Taylor & Francis Group.

Sanna, L. J., Chang, E. C., Miceli, P. M., & Lundberg, K. B. (2011). Rising up to higher virtues: Experiencing elevated physical height uplifts prosocial actions. *Journal of Experimental Social Psychology, 47,* 472–476. [Retracted article]

Saris, W. E., & Gallhofer, I. N. (2007). *Design, evaluation, and analysis of questionnaires for survey research.* Hoboken, NJ: Wiley.

Schafer, J. L., & Graham, J. W. (2002). Missing data: Our view of the state of the art. *Psychological Methods, 7,* 147–177. https://doi.org/10.1037//1082-989X.7.2.147

Seaman, J. W. Jr., Odell, P. S., & Young, D. M. (1985). Maximum variance unimodal distributions. *Statistics & Probability Letters, 3*, 255–260.

Shadish, W. R., Cook, T. D., & Campbell, D. T. (2002). *Experimental and quasi-experimental designs for generalized causal inference, Volume 1*. Boston, MA: Houghton Mifflin.

Sijtsma, K. (2016). Playing with data—Or how to discourage questionable research practices and stimulate researchers to do things right. *Psychometrika, 81*, 1–15.

Sijtsma, K., Emons, W. H. M., Steneck, N. H., & Bouter, L. M. (2021). Steps toward preregistration of research on research integrity. *Research Integrity and Peer Review, 6*. https://doi.org/10.1186/s41073-021-00108-4

Sijtsma, K., & Van der Ark, L. A. (2021). *Measurement models for psychological attributes*. Boca Raton, FL: Chapman & Hall/CRC.

Simmons, J. P., Nelson, L. D., & Simonsohn, U. (2011). False-positive psychology: Undisclosed flexibility in data collection and analysis allows presenting anything as significant. *Psychological Science, 22*, 1359–1366. https://doi.org/10.1177/0956797611417632

Simonsohn, U. (2013). Just post it: The lesson from two cases of fabricated data detected by statistics alone. *Psychological Science, 24*, 1875–1888. https://doi.org/10.1177/0956797613480366

Smit, M. (2012). De val van Don Poldermans (Don Poldermans' downfall). *Medisch Contact, 67*, 874–878.

Stapel, D. A., & Lindenberg, S. (2011). Coping with chaos: How disordered contexts promote stereotyping and discrimination. *Science, 332*, 251–253. [Retracted article]

Steneck, N. H. (2006). Fostering integrity in research: Definitions, current knowledge, and future directions. *Science and Engineering Ethics, 12*, 53–74.

Stricker, J., & Günther, A. (2019). Scientific misconduct in psychology. A systematic review of prevalence estimates and new empirical data. *Zeitschrift für Psychologie, 227*, 53–63. https://doi.org/10.1027/2151-2604/a000356

Stroebe, W. (2019). What can we learn from many labs replications? *Basic and Applied Social Psychology, 41*, 91–103. https://doi.org/10.1080/01973533.2019.1577736

Stroebe, W., Postmes, T., & Spears, R. (2012). Scientific misconduct and the myth of self-correction in science. *Perspectives on Psychological Science, 7*, 670–688. https://doi.org/10.1177/1745691612460687

Tabachnik, B. G., & Fidell, L. S. (2007). *Using multivariate statistics*. Boston, MA: Pearson.

Thavarajah, S., White, W. B., & Mansoor, G. A. (2003). Terminal digit bias in a specialty hypertension faculty practice. *Journal of Human Hypertension, 17*, 819–822. Downloaded from: https://www.nature.com/articles/1001625.pdf?origin=ppub

Trafimow, D. (2018). An a priori solution to the replication crisis. *Philosophical Psychology, 31*(8), 1188–1214. https://doi.org/10.1080/09515089.2018.1490707

Trafimow, D., Amrhein, V., Areshenkoff, C. N., Barrera-Causil, C. J., Beh, E. J. et al. (2018). Manipulating the alpha level cannot cure significance testing. *Frontiers in Psychology*. doi:10.3389/fpsyg.2018.00699

Tsuruda, K. M., Hofvind, S., Akslen, L. A., Hoff, S. R., & Veierød. M. B. (2020). Terminal digit preference: A source of measurement error in breast cancer diameter reporting. *Acta Oncologica, 59*, 260–267. https://doi.org/10.1080/0284186X.2019.1669817

Tversky, A., & Kahneman, D. (1971). Belief in the law of small numbers. *Psychological Bulletin, 76*, 105–110.

Tversky, A., & Kahneman, D. (1973). Availability: A heuristic for judging frequency and probability. *Cognitive Psychology, 5*, 207–232.

Tversky, A., & Kahneman, D. (1974). Judgment under uncertainty: Heuristics and biases. *Science, 185*, 1124–1131.

Ulrich, R., & Miller, J. (2020). Questionable research practices may have little effect on replicability. *eLife, 9*, e58237. https://doi.org/10.7554/eLife.58237

Van Belle, G. (2008). *Statistical rules of thumb*. Hoboken, NJ: Wiley.

Van Buuren, S. (2018). *Flexible imputation of missing data*. Boca Raton, FL: Chapman & Hall/CRC.

Van de Schoot, R., Kaplan, D., Denissen, J., Asendorpf, J. B., Neyer et al. (2014). A gentle introduction to Bayesian analysis: Applications to developmental research. *Child Development, 85*, 842–860.

Van de Schoot, R., Winter, S. D., Griffioen, E., Grimmelikhuijsen, S, Arts, I. et al. (2021). The use of questionable research practices to survive in academia examined with expert elicitation, prior-data conflicts, Bayes factors for replication effects, and the Bayes truth serum. *Frontiers in Psychology, Quantitative Psychology and Measurement, 12*, 621547. https://doi.org/10.3389/fpsyg.2021.621547

Van den Berg, R. (2021). *Een gedreven buitenstaander. J. H. van 't Hoff, de eerste Nobelprijswinnaar voor scheikunde (An enthusiastic outsider. J. H. van 't Hoff, first Nobel laureate for chemistry)*. Amsterdam: Prometheus.

Van Dijk, W., Ouwerkerk, J., & Vliek, M. (2015). *Verslag ASPO-commissie: bijdragen Diederik Stapel aan het Jaarboek Sociale Psychologie.* ASPO Press. http://www.sociale-psychologie.nl/wpcontent/uploads/2015/06/Verslag-ASPO-commissie-van-Dijk.pdf

Van Doorn, J., Van den Bergh, D., Böhm, U., Dablander, F., Derks, K. et al. (2021). The JASP guidelines for conducting and reporting a Bayesian analysis. *Psychonomic Bulletin and Review, 28,* 813–826. https://doi.org/10.3758/s13423-020-01798-5

Van Ginkel, J. R., Sijtsma, K., Van der Ark, L. A., & Vermunt, J. K. (2010). Incidence of missing item scores in personality measurement, and simple item-score imputation. *Methodology: European Journal of Research Methods for the Behavioral and Social Sciences, 6,* 17–30.

Van Kolfschoten, F. (1993). *Valse vooruitgang. Bedrog in de Nederlandse wetenschap (False progress. Deceit in Dutch science).* Amsterdam: Uitgeverij Contact.

Van Kolfschoten, F. (2012). *Ontspoorde wetenschap. Over fraude, plagiaat en academische mores (Derailed science. About fraud, plagiarism and academic standards).* Amsterdam: Uitgeverij De Kring.

Vines, T. H., Albert, A. Y. K., Andrew, R. L., Débarre, F., Bock, D. G. et al. (2014). The availability of research data declines rapidly with article age. *Current Biology, 24,* 94–97. http://dx.doi.org/10.1016/j.cub.2013.11.014

Wagenmakers, E. J. (2007). A practical solution to the pervasive problems of p values. *Psychonomic Bulletin & Review, 14,* 779–804.

Wagenmakers, E. J., Marsman, M., Jamil, T., Ly, A., & Verhagen, J. (2018). Bayesian inference for psychology: Part I: Theoretical advantages and practical ramifications. *Psychonomic Bulletin &Review, 25,* 35–57. https://doi.org/10.3758/s13423-020-01798-5

Wagenmakers, E. J., Wetzels, R., Borsboom, D., & Van der Maas, H. L. J. (2011). Why psychologists must change the way they analyze their data: The case of Psi: Comment on Bem (2011). *Journal of Personality and Social Psychology, 100,* 426–432. https://doi.org/10.1037/a0022790

Wagenmakers, E. J., Wetzels, R., Borsboom, D., Van der Maas, H. L. J., & Kievit, R. A. (2012). An agenda for purely confirmatory research. *Perspectives on Psychological Science, 7,* 632–638.

Wainer, H. (2016). *Truth or truthiness. Distinguishing fact from fiction by learning to think like a datascientist.* Cambridge: Cambridge University Press.

Wallach, J. D., Boyack, K. W., & Ioannidis, J. P. A. (2018). Reproducible research practices, transparency, and open access data in the biomedical literature, 2015–2017. *PLoS Biology*, *16*(11), e2006930. https://doi.org/10.1371/journal.pbio.2006930

Wicherts, J. M., & Bakker, M. (2012). Publish (your data) or (let the data) perish! Why not publish your data too? *Intelligence*, *40*, 73–76.

Wicherts, J. M., Borsboom, D., Kats, J., & Molenaar, D. (2006). The poor availability of psychological research data for reanalysis. *American Psychologist*, *61*, 726–728.

Wicherts, J. M., Veldkamp, C. L. S., Augusteijn, H. E. M., Bakker, M., Van Aert, R. C. M. et al. (2016). Degrees of freedom in planning, running, analyzing, and reporting psychological studies: A checklist to avoid *p*-hacking. *Frontiers in Psychology*, *7*. https://doi.org/10.3389/fpsyg.2016.01832

Wiggins, J. S. (1973). *Personality and prediction. Principles of personality assessment.* Menlo Park, CA: Addison-Wesley.

Wild, C. J., & Pfannkuch, M. (1999). Statistical thinking in empirical enquiry (with discussion). *International Statistical Review*, *67*, 223–265.

Wilkinson, M. D., Dumontier, M., Aaldersberg, I. J. J., Appleton, G., Axton, M. et al. (2016). The FAIR Guiding Principles for scientific data management and stewardship. *Scientific Data*, 3, 160018. https://doi.org/10.1038/sdata.2016.18

Winer, B. J. (1971). *Statistical principles in experimental design.* Tokyo: Mc-Graw-Hill.

Zwart, H. (2017). *Tales of research misconduct. A Lacanian diagnostics of integrity challenges in science novels.* Cham: Springer Nature.

Index

Pages in *italics* refer figures, pages in **bold** refer tables, and pages followed by n refer notes.